Name _____

 **Take Your Pla**

Write the place and value for each underlined digit.

A. 2,351   thousands   2,000 _____

B. 43,920 _____

C. 592,103 _____

D. 864,135 _____

E. 972,468 _____

F. 237,904 _____

Match each number to its expanded form.

G. 342,179   900,000 + 60,000 + 3,000 + 400 + 70 + 1

H. 567,804   300,000 + 40,000 + 2,000 + 100 + 70 + 9

I. 963,471   500,000 + 7,000 + 800 + 40

J. 903,481   500,000 + 60,000 + 7,000 + 800 + 4

K. 507,840   900,000 + 60,000 + 3,000 + 400 + 1

L. 963,401   900,000 + 3,000 + 400 + 80 + 1

M. 57,084   90,000 + 3,000 + 400 + 70 + 1

N. 93,471   500,000 + 60,000 + 7,000 + 800

O. 567,800   50,000 + 7,000 + 80 + 4

Name _____ Place value

Place Value Village

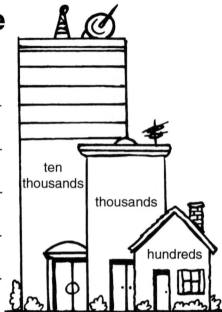

Write the place and the value of the digit 7 in each number.

A. 5,713   hundreds   700

B. 7,034   _____

C. 793,400 _____

D. 375,920 _____

E. 791,468 _____

Write the numbers.

F. two thousand, eight hundred fifty-seven   2,857

G. fifty-one thousand, five hundred twelve   _____

H. 3,000 + 400 + 70 + 6   _____

I. 20,000 + 8,000 + 900 + 40 + 1   _____

J. six hundred eleven thousand, four hundred one   _____

K. nine hundred fifty-six thousand, forty-seven   _____

L. 800,000 + 70,000 + 6,000 + 500 + 40 + 3   _____

M. 900,000 + 4,000 + 300 + 7   _____

N. one hundred three thousand, eight hundred twelve   _____

O. ninety-seven thousand, three hundred twenty   _____

P. 700,000 + 4,000 + 700 + 4   _____

© Frank Schaffer Publications, Inc.   2   FS-32071 Fourth Grade Math Review

Name _____ Place value

Millions of Years Ago

Write each number.

A. six million, three hundred two thousand, forty-two   6,302,042

B. forty-three million, three hundred twenty thousand, six hundred   _____

C. six hundred million, forty-three thousand, two hundred one   _____

D. five hundred sixteen million, three hundred twelve thousand   _____

E. two hundred five million, three hundred twenty-one thousand   _____

F. twenty-five million, two hundred thousand, six hundred fifty   _____

G. six hundred million, forty-three thousand, three hundred twenty-one   _____

H. five hundred one million, three thousand, eight hundred ninety-six   _____

I. twenty-five million, six hundred fifty-two   _____

J. sixty million, one hundred forty-three thousand, two hundred   _____

K. fifty-six million, one hundred ten thousand, one   _____

L. four hundred thirty million, three hundred twenty thousand, six   _____

M. four hundred twenty-five million, three hundred   _____

Study the number **425,368,197**. Write the digit and the value for each place listed below.

N. tens   9   90

O. hundreds   _____

P. thousands   _____

Q. ones   _____

R. millions   _____

S. ten millions _____

T. hundred thousands _____

U. hundred millions _____

V. ten thousands _____

© Frank Schaffer Publications, Inc.   3   FS-32071 Fourth Grade Math Review

Name _____ Place value

Number the Stars

Write the digit and value for the given places in **817,623,954**.

A. tens ____5____ _____50_____

B. thousands _____

C. hundred thousands _____

D. ten millions _____

E. hundreds _____

F. ones _____

G. millions _____

H. hundred millions _____

I. ten thousands _____

Write each number.

J. five million, six hundred eight thousand   5,608,000

K. two hundred million, five hundred thousand   _____

L. nine million, forty-three thousand, fifteen   _____

M. eight hundred fifty-eight million   _____

N. one hundred six million, four hundred thousand   _____

O. seven hundred eleven million, ninety thousand   _____

P. sixty-one million, nine hundred thousand, two   _____

Q. four hundred seventy-eight million, thirteen   _____

R. nine hundred two million, seventy-six thousand   _____

S. one million, two hundred thousand, three hundred   _____

T. six million, eight hundred fifty-three thousand   _____

U. seventeen million, five hundred thousand   _____

V. four hundred eighty-two million, one hundred six   _____

W. nine hundred million, nine thousand, ninety   _____

© Frank Schaffer Publications, Inc.   4   FS-32071 Fourth Grade Math Review

Name _____   Comparing and ordering numbers

You're the Greatest!

Compare each pair of numbers. Write < or > in each ◯.

A.   5,361   ⊗   5,300

B.   9,327   ◯   9,237

C.   10,567   ◯   10,651   235,400   ◯   234,900

D.   593,461   ◯   593,614   893,982   ◯   892,983

E.   98,997   ◯   100,016   203,960   ◯   230,141

F.   497,843   ◯   496,912   156,651   ◯   651,156

G.   600,000   ◯   600,010   93,825   ◯   94,053

H.   1,493,017   ◯   1,947,413   23,985,310   ◯   29,385,013

I.   31,113,311   ◯   13,331,113   19,675,902   ◯   20,001,000

J.   1,087,789   ◯   987,911   63,814,910   ◯   59,950,418

Arrange the numbers in each group from the least to the greatest.

K.   59,359;   590,359;   509,359;   95,359

_____

L.   417,003;   950,398;   409,985;   398,051

_____

M.   24,890;   20,561;   24,279;   24,385

_____

N.   831,485;   813,485;   89,497;   830,549

_____

© Frank Schaffer Publications, Inc.   5   FS-32071 Fourth Grade Math Review

Name _____   Comparing and ordering numbers

Order in the Court

Compare the numbers in each pair. Write < or > in the ☐.

A.   8,685   [<]   8,698

B.   29,402   ☐   29,420   38,822   ☐   32,988

C.   83,920   ☐   89,320   153,917 ☐   149,395

D.   475,315 ☐   475,309   893,492 ☐   893,489

E.   924,517 ☐   924,496   267,103 ☐   267,130

F.   1,000,310 ☐ 998,897   1,234,511 ☐ 1,235,490

G.   56,375,011 ☐ 53,983,114   4,753,980 ☐ 4,754,001

H.   50,800,114 ☐ 49,990,099   9,837,410 ☐ 10,000,000

Order from the least to the greatest.

I.   13,457; 24,537; 23,745; 14,753

_____

J.   46,842; 47,428; 46,488; 47,802

_____

K.   181,850; 180,058; 182,148; 189,984

_____

L.   289,416; 191,989; 188,063; 290,104

_____

M.   197,963; 202,450; 200,001; 199,982

_____

© Frank Schaffer Publications, Inc.   6   FS-32071 Fourth Grade Math Review

Name _____ Rounding

Merry-Go-Round

Round each number to the nearest thousand.

A. 2,458 _____2,000_____

B. 11,592 _____   25,923 _____

C. 172,461 _____   321,390 _____

D. 493,852 _____   825,495 _____

E. 515,151 _____   936,683 _____

Round each number to the nearest ten thousand.

F. 36,198 _____   14,572 _____

G. 322,039 _____   654,567 _____

H. 167,457 _____   89,189 _____

I. 351,956 _____   231,097 _____

J. 154,631 _____   458,532 _____

Round each number to the nearest hundred thousand.

K. 169,842 _____   562,719 _____

L. 452,008 _____   346,135 _____

M. 351,637 _____   239,658 _____

N. 775,693 _____   315,521 _____

O. 987,463 _____   1,405,929 _____

Name _____   Rounding

Round and Round She Goes

Round to the nearest thousand.

A. 4,357   __4,000__

B. 9,093   _____

C. 6,572   _____   3,675   _____

D. 27,772   _____   41,056 _____

E. 86,915   _____   42,411 _____

Round to the nearest ten thousand.

F. 67,497   _____   15,926 _____

G. 84,675   _____   44,893 _____

H. 91,842   _____   53,492 _____

I. 318,950 _____   128,147 _____

J. 596,715 _____   231,849 _____

Round to the places indicated on the chart below.

	Numbers	nearest thousand	nearest ten thousand	nearest hundred thousand
K.	143,987			
L.	562,490			
M.	275,856			
N.	2,961,758			
O.	1,425,897			

Name _____   Addition and subtraction facts

   **Fast-track Facts**

Add or subtract.

A.	2 + 9	6 + 7	9 − 8	14 − 7	6 + 5	7 + 9
B.	16 − 8	9 + 4	7 + 7	17 − 9	11 − 8	2 + 9
C.	3 + 9	15 − 8	7 + 3	4 + 8	9 + 8	10 − 5
D.	13 − 8	12 − 9	4 + 6	15 − 6	12 − 7	13 − 6

E. 18 − 9 = _____   14 − 8 = _____   7 + 8 = _____

F. 14 − 5 = _____   6 + 9 = _____   5 + 7 = _____

G. 16 − 9 = _____   7 + 4 = _____   8 + 8 = _____

H. 17 − 8 = _____   9 + 5 = _____   8 + 6 = _____

I. 8 + 4 = _____   15 − 9 = _____

J. 16 − 7 = _____   5 + 8 = _____

K. 9 + 9 = _____   13 − 4 = _____

© Frank Schaffer Publications, Inc.   9   FS-32071 Fourth Grade Math Review

Name _____   Addition and subtraction facts

Just the Facts

Find each sum or difference. Watch the signs!

A. 9 + 8 = _____   10 − 4 = _____   12 − 9 = _____

B. 16 − 8 = _____   5 + 8 = _____   6 + 6 = _____

C. 15 − 6 = _____   7 + 9 = _____   15 − 7 = _____

D. 12 − 6 = _____   9 + 2 = _____   8 + 6 = _____

E. 13 − 4 = _____   8 + 8 = _____   16 − 9 = _____

F. 6 + 7 = _____   10 − 2 = _____   12 − 3 = _____

G. 12 − 6 = _____   13 − 8 = _____   14 − 6 = _____

H.   4   6   11   10   7   18
   + 7   + 4   − 9   − 9   + 8   − 9

I.   5   14   11   12   3   2
   + 5   − 7   − 8   − 5   + 9   + 9

J.   9   6   10   15
   + 9   + 8   − 3   − 9

K.   14   7   5   6
   − 8   + 8   + 7   + 9

© Frank Schaffer Publications, Inc.

Name _____ Estimating sums and differences

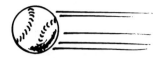 # In the Ballpark

Round each number to its highest place. Then estimate the sums and differences.

A. 61 ∘∘ (rounds to 60)   92   41   60
   + 73 ∘∘ (rounds to 70) − 14   + 49   − 48
    130 ∘∘ (60 + 70 = 130)

B.   378   603   729   746
    − 103   − 485   + 196   + 718

C.   927   842   903   296
    + 896   − 645   − 875   + 849

D. 6,899   2,985   987   3,795
   − 2,463   + 3,109   + 4,106   − 1,128

E. 7,537   9,205   8,831   7,015
   + 2,486   − 6,886   + 5,679   − 5,983

F. 23,857   96,810   47,923   83,519
   + 19,452 − 28,563   + 13,526   − 30,274

© Frank Schaffer Publications, Inc.   11   FS-32071 Fourth Grade Math Review

Name _____   Estimating sums and differences

In the Rough

Round each number to its highest place. Then estimate the sums.

A. 64 + 43 = ⟨60 + 40 = 100⟩ 100   78 + 52 = _____

B. 51 + 36 = _____   85 + 72 = _____

C. 381 + 278 = _____   144 + 585 = _____

D. 562 + 897 = _____   659 + 845 = _____

E. 2,965 + 3,149 = _____   1,782 + 3,946 = _____

F. 6,854 + 7,310 = _____   8,483 + 4,701 = _____

G. 29,357 + 31,468 = _____   65,826 + 12,419 = _____

H. 43,925 + 28,046 = _____   81,997 + 8,692 = _____

Round each number to its highest place. Then estimate the differences.

I. 78 − 54 = _____   53 − 29 = _____

J. 69 − 61 = _____   92 − 89 = _____

K. 493 − 147 = _____   525 − 237 = _____

L. 263 − 98 = _____   807 − 485 = _____

M. 5,398 − 2,427 = _____   9,872 − 5,188 = _____

N. 8,312 − 7,903 = _____   3,786 − 2,241 = _____

O. 78,845 − 21,341 = _____   94,378 − 52,481 = _____

P. 56,012 − 39,085 = _____   26,834 − 16,296 = _____

© Frank Schaffer Publications, Inc.   FS-32071 Fourth Grade Math Review

Name _____ Adding whole numbers

"Sum"mer Fun!

Add. Circle each sum in the number box at the bottom of the page.

A.
   ¹93   47   37   86
+ 69   + 84   + 55   + 47
_162_

B.
  37   38   26   54   69
+ 46   + 73   + 26   + 75   + 84

C.
  135   247   530   547   448
+ 36   + 161   + 193   + 296   + 539

D.
  296   620   322   123   302
  347   66   497   546   184
+ 25   + 137   + 86   + 97   + 436

(1)	3	1	4	7	7	1	4	8	3
6	0	0	2	0	0	5	4	9	6
(2)	0	0	1	2	9	3	7	2	7
9	2	0	7	8	2	3	2	2	9
6	1	3	3	0	0	9	3	1	8
6	4	5	6	5	1	0	0	0	7
8	0	2	0	2	7	5	8	4	3
4	0	8	1	1	1	7	6	6	0

© Frank Schaffer Publications, Inc.   13   FS-32071 Fourth Grade Math Review

Name _____ Adding whole numbers

Across and Down, Around the Town

Find the sum for each problem. Write the sums on the crossword puzzle grid.

Across
A. 631 + 731 = __1,362__
D. 85 + 383 + 99 = _____
G. 375 + 28 = _____
I. 32 + 56 = _____
J. 27 + 12 + 26 = _____
L. 19 + 108 = _____
N. 23 + 14 + 19 + 18 = _____
P. 13 + 24 = _____
S. 255 + 36 + 108 = _____
V. 25 + 16 + 8 = _____
X. 32 + 48 = _____
Y. 429 + 350 + 23 = _____

Down
A. 130 + 16 = _____
B. 55 + 9 = _____
C. 99 + 49 + 53 = _____
E. 52 + 16 = _____
F. 731 + 49 = _____
H. 25 + 7 = _____
K. 38 + 19 = _____
M. 28 + 45 = _____
O. 11 + 32 = _____
Q. 54 + 20 = _____
R. 16 + 29 + 43 = _____
T. 49 + 49 = _____
U. 73 + 17 = _____
W. 26 + 50 + 22 = _____

© Frank Schaffer Publications, Inc.   14   FS-32071 Fourth Grade Math Review

Name _____ Adding whole numbers

Totally Cool

Find the sums.

A.   2,345   4,614   2,937   5,807
      + 3,872   + 3,747   + 1,758   +   956
       6,217

B.   6,821   2,436   2,385   5,661
      +   59   + 3,827   + 3,561   + 3,359

C.   1,986   2,931   4,385   5,188
      + 5,923   + 3,142   + 3,579   + 1,996

D.   352   2,481   3,105   4,385
      + 7,863   + 6,879   + 4,896   + 2,961

E. $ 35.19 $ 68.75 $ 43.50 $ 38.46
      + 26.85   +   9.97   + 29.79   + 17.38

F.   3,125   5,687 $ 23.95
        936   47   8.97
      + 1,215   +   863   + 15.60

Name_____ Adding whole numbers

Cashier's Delight

Add.

A. 4,382   5,783   1,983   2,520
   + 1,166   + 927   + 1,872   + 3,980
   ─────
   5,548

B. 7,193   1,735   2,961   896
   + 1,875   + 3,961   + 1,446   + 8,473

C. 6,388   3,871   8,375   5,309
   + 1,793   + 5,619   + 924   + 3,998

D. 4,986   9,758   8,234   3,685
   + 1,194   + 162   + 1,398   + 4,839

E. $ 20.64 $ 59.41 $ 26.53 $ 89.95
   + 53.47   + 38.64   + 61.74   + 9.97

F. 2,357   4,285 $ 15.65
    3,496   1,846   8.27
   + 196   + 2,099   + 42.93

16

© Frank Schaffer Publications, Inc.   FS-32071 Fourth Grade Math Review

Name _____ Adding whole numbers

Larger-Than-Life

Find the sums.

A. 32,967   63,937   15,816   89,471
   + 19,824   + 4,852   + 9,537   + 3,895
   ─────────
    52,791

B. 23,689   25,963   43,895   56,140
   + 18,497   + 17,827   + 16,742   + 19,396

C. 27,975   56,910   25,926   18,888
   + 38,046   + 9,438   + 18,409   + 47,196

D. $ 384.16 $ 109.83 $ 258.32 $ 585.46
   +   19.81   + 120.79   + 570.07   +   18.54

E. 28,532   58,896   67,242
     1,829   1,746   2,037
   + 32,047   + 25,814   +15,384

F. $ 199.36 $ 89.74 $ 356.89
     270.46   146.75   28.34
   +   89.95   + 689.61   + 563.57

© Frank Schaffer Publications, Inc.   FS-32071 Fourth Grade Math Review

Name _____ Adding whole numbers

The Great One

Add.

A. 19,346   26,128   9,286   83,126
   + 28,981   + 58,642   + 36,697   + 8,567
   ────────
    48,327

B. 11,987   43,826   56,046   46,916
   + 8,976   + 10,897   + 31,965   + 28,427

C. 25,896   39,126   41,830   75,326
   + 24,104   + 22,198   + 9,267   + 19,937

D. $ 238.42 $ 463.02 $ 379.16 $ 538.36
   + 162.46   + 189.98   + 438.09   + 91.07

E. 28,357   89,432 $ 258.36 $ 398.76
     1,963   987   192.46   18.07
   + 46,107   + 6,792   + 98.10   + 419.63

F. 38,916 $ 821.63
    43,019   119.63
   + 7,923   + 57.96

© Frank Schaffer Publications, Inc.   18   FS-32071 Fourth Grade Math Review

Name _____ Subtracting whole numbers

Centipede Subtraction

Subtract. Then use the code to solve the riddle.

A 1̶3̶1 B 523 C 946 D   95 E 927
  −124   −196   −832   − 77   −483
    7

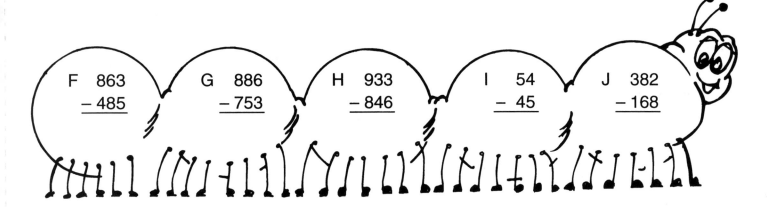

F 863   G 886   H 933   I   54   J 382
  −485   −753   −846   − 45   −168

K 450   L 250   M 574   N   82   O 411
  −365   −249   −468   − 58   −187

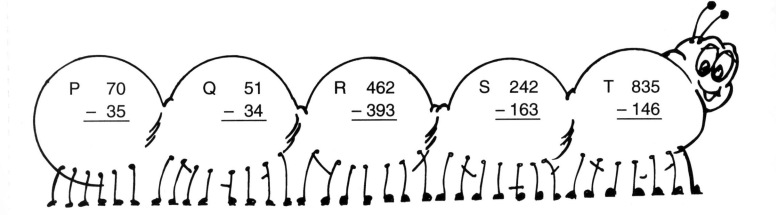

P   70   Q   51   R 462   S 242   T 835
  − 35   − 34   −393   −163   −146

What did the centipede say to its mother?

___ ___ ___ ___   ___ ___ A ,
 1   224 224   85   106   7

___ ___   ___ ___ A ___ ___ ___ !
24   224   87   7   24   18   79

Name _____ Subtracting whole numbers

Missing in Action

Find the missing digits.

A.
```
   5 [2]
 - 3 9
 ─────
   1 3
```
```
   5 [ ] 6
 - 1 9 8
 ───────
   3 9 8
```
```
   6 2 [ ]
 -   1 1
 ───────
   6 1 0
```
```
   4 9 6
 - 2 5 [ ]
 ───────
   2 4 1
```

B.
```
   8 3
 - [ ] 1
 ─────
   4 2
```
```
   4 5 [ ]
 - 2 6 5
 ───────
   1 8 5
```
```
   3 3 2
 - [ ] 7
 ───────
   2 6 5
```
```
   4 7 3
 - [ ] 2
 ───────
   3 8 1
```

C.
```
   6 8
 - 4 [ ]
 ─────
   1 9
```
```
   9 1
 - [ ] 3
 ─────
   2 8
```
```
   7 7 [ ]
 -   1 7
 ───────
   7 5 7
```
```
   5 [ ] 2
 - 1 6 6
 ───────
   3 6 6
```

D.
```
   3 3 [ ]
 - 1 4 8
 ───────
   1 8 4
```
```
   3 5 3
 - 1 [ ] 8
 ───────
   1 7 5
```
```
   8 9 [ ]
 - 1 6 7
 ───────
   7 2 8
```
```
   9 1 2
 - 4 6 [ ]
 ───────
   4 4 6
```

E.
```
   2 [ ] 0
 - 1 7 5
 ───────
     7 5
```
```
   9 [ ]
 - 3 6
 ─────
   5 4
```
```
   8 5 3
 - [ ] 4
 ───────
   7 6 9
```

F.
```
   6 [ ] 3
 -   3 7
 ───────
   6 2 6
```
```
   8 2 1
 - 2 [ ] 6
 ───────
   6 1 5
```
```
   1 9 [ ]
 - 1 2 5
 ───────
     6 8
```

Name _____ Subtracting whole numbers

Leftovers

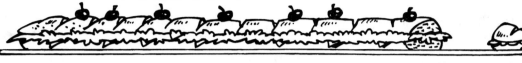

Write the differences.

A.
```
  4 11 1
  5,237        6,895        1,596        6,735
- 1,856      - 2,167      -   898      - 1,948
  -----
  3,381
```

B.
```
  4,327        7,280        1,244        3,850
- 1,175      - 5,461      -   922      - 3,760
```

C.
```
  5,556        4,392        6,432        5,324
- 2,658      - 1,774      -   265      - 1,427
```

D.
```
  6,442        6,314        8,735        4,744
- 2,795      - 5,719      - 5,787      - 3,658
```

E.
```
  $ 42.38      $ 17.71      $ 92.63      $ 15.35
-    9.39    -    2.88    -   65.94    -   13.69
```

F.
```
  $ 62.08      $ 68.25      $ 72.56      $ 24.82
-   38.52    -    4.09    -   61.84    -   18.65
```

© Frank Schaffer Publications, Inc. 21 FS-32071 Fourth Grade Math Review

Name _____   Subtracting whole numbers

All That's Left

Find the differences.

A.
    3 14 1
    4,~~5~~29
 − 1,635
    2,894

    6,218
 − 3,862

    9,126
 − 7,241

    7,843
 − 3,589

B.
    8,942
 − 1,385

    1,549
 −   425

    6,961
 − 4,682

    3,257
 − 3,098

C.
    8,573
 − 2,791

    9,836
 − 1,465

    3,816
 −   942

    8,414
 − 3,916

D.
    3,715
 − 1,896

    4,536
 − 2,718

    6,448
 − 4,942

    1,815
 −   927

E.
  $ 15.95
 −   9.98

  $ 53.10
 − 49.95

  $ 87.50
 − 49.25

F.
  $ 26.27
 − 19.35

  $ 45.11
 − 28.50

  $ 78.76
 − 59.47

© Frank Schaffer Publications, Inc.   22   FS-32071 Fourth Grade Math Review

Name _____ Subtracting across zeros

Across the River

Subtract.

A. $\overset{2\,\,9\,\,1}{\cancel{3}\cancel{0}0}$ 805 200
    − 125   − 49 − 164
    ───   ───   ─────
    175

B. 502   4,307   5,900   5,302
    − 159   − 439   − 1,375   − 1,953

C. 401   1,201   5,002   3,000
    − 395   − 986   − 2,005   − 889

D. 6,004   5,089   5,024   4,702
    − 4,832   − 498   − 1,954   − 1,794

E. $ 6.05 $ 7.07 $ 9.04 $ 4.00
    − 5.87   − 2.38   − 5.78   − 0.66

F. $ 58.05 $ 20.05 $ 31.06 $ 80.05
    − 19.98   − 9.87   − 13.67   − 76.19

© Frank Schaffer Publications, Inc.   23   FS-32071 Fourth Grade Math Review

Name _____   Subtracting across zeros

Leapfrog

Subtract.

A. ²3̸0̸⁹0̸¹   100   402   700
    − 174   − 53   − 394   − 492
    ─────
     126

B.   503   400   3,200   4,701
   − 147   − 278 − 1,695 − 2,382

C. 1,702   2,903   6,001   4,067
   − 375 − 1,594 − 5,873 − 1,970

D. 2,008   5,100   1,000   9,022
   − 732 − 1,163   − 973 − 6,375

E. $ 8.01 $ 3.01 $ 80.40 $ 14.06
   − 2.32   − 0.67   − 23.63   − 5.98

F. $ 10.00 $ 23.01
    − 4.97   − 18.75

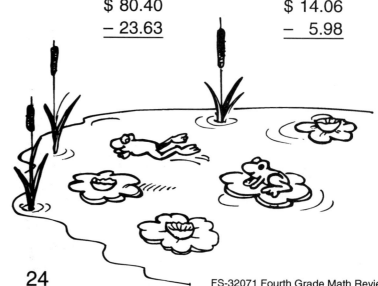

© Frank Schaffer Publications, Inc.   24   FS-32071 Fourth Grade Math Review

Name _____ Adding to check subtraction

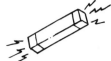 

Opposites Attract

Find each difference. Check your work by adding.

A. $\begin{array}{r}4\,10\,1\\ \cancel{5}\cancel{1}3\\ -285\\ \hline 228\end{array}$ → $\begin{array}{r}228\\ +285\\ \hline 513\end{array}$

B. $\begin{array}{r}830\\ -497\\ \hline \end{array}$

C. $\begin{array}{r}903\\ -886\\ \hline \end{array}$

D. $\begin{array}{r}700\\ -196\\ \hline \end{array}$

E. $\begin{array}{r}1{,}467\\ -835\\ \hline \end{array}$

F. $\begin{array}{r}2{,}385\\ -1{,}426\\ \hline \end{array}$

G. $\begin{array}{r}35{,}801\\ -19{,}625\\ \hline \end{array}$

H. $\begin{array}{r}27{,}536\\ -12{,}147\\ \hline \end{array}$

I. $\begin{array}{r}\$\,5.00\\ -3.79\\ \hline \end{array}$

J. $\begin{array}{r}\$\,41.35\\ -9.98\\ \hline \end{array}$

K. $\begin{array}{r}\$\,235.09\\ -83.60\\ \hline \end{array}$

L. $\begin{array}{r}\$\,545.02\\ -359.17\\ \hline \end{array}$

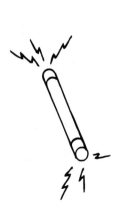

© Frank Schaffer Publications, Inc. FS-32071 Fourth Grade Math Review

Name _____   Adding to check subtraction

Checkmate!

Subtract. Then add to check.

A. 7 9 1
    8̸0̸7   288   935   845
    − 519   + 519   − 413   − 790
    ‾‾‾‾‾   ‾‾‾‾‾
    288   807

B. 702   657   915
    − 416   − 95   − 385

C. $ 2.79 $ 9.00 $ 4.03
    − 1.85   − 3.87   − 1.86

D. 5,375   9,003
    − 2,832   − 6,894

E. $ 68.75 $ 83.00
    − 32.59   − 69.25

F. 23,897 $ 630.03
    − 18,261   − 281.55

Name _____   Multiplication facts

Pop a Wheelie

Find the products.

A.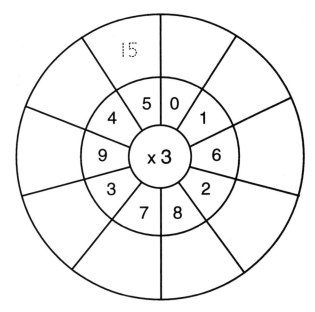

B.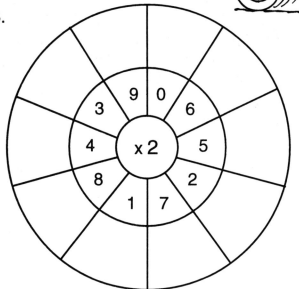

C.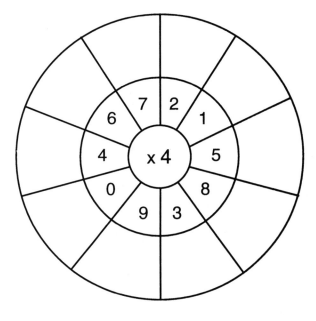

D.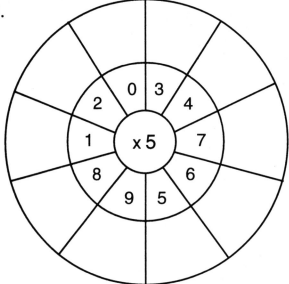

E.   0   3   5   2   0   5   4   9
    x 9   x 7   x 0   x 8   x 7   x 6   x 8   x 3

Name _____   Multiplication facts

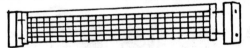   # Table Tennis

Complete each multiplication table.

A.

	x 2
3	6
7	
5	
9	
4	
1	
6	
2	
0	
8	

	x 3
3	
0	
4	
1	
6	
5	
8	
7	
9	
2	

	x 4
5	
2	
0	
1	
6	
9	
3	
8	
4	
7	

	x 5
2	
1	
5	
9	
0	
6	
8	
4	
7	
3	

Write the products.

B. 3 x 0 = _____   1 x 4 = _____   2 x 7 = _____

C. 0 x 5 = _____   6 x 1 = _____   3 x 8 = _____

D. 4 x 3 = _____   2 x 0 = _____   5 x 1 = _____

E. 5 x 2 = _____   0 x 6 = _____   1 x 8 = _____

F. 1 x 7 = _____   8 x 0 = _____   9 x 2 = _____

G. 2 x 8 = _____   4 x 6 = _____   5 x 4 = _____

H. 9 x 1 = _____   7 x 0 = _____   4 x 9 = _____

I. 5 x 5 = _____   2 x 4 = _____   0 x 1 = _____

J. 1 x 1 = _____   0 x 0 = _____   6 x 5 = _____

© Frank Schaffer Publications, Inc.   28   FS-32071 Fourth Grade Math Review

Name _____   Multiplication facts

Fast Facts

Write the products.

A. 1 x 6 = _____   5 x 8 = _____   3 x 9 = _____

B. 4 x 7 = _____   2 x 6 = _____   5 x 7 = _____

C. 6 x 9 = _____   7 x 8 = _____   5 x 9 = _____

D. 2 x 7 = _____   3 x 6 = _____   1 x 9 = _____

E. 4 x 9 = _____   4 x 8 = _____   7 x 9 = _____

F. 8 x 8 = _____   0 x 9 = _____   9 x 8 = _____

G. 4 x 6 = _____   2 x 9 = _____   9 x 3 = _____

H. 7 x 4 = _____   9 x 4 = _____   9 x 7 = _____

I. 6 x 8 = _____   3 x 8 = _____   0 x 8 = _____

J. 1 x 8 = _____   0 x 6 = _____   5 x 6 = _____

K. 0 x 7 = _____   7 x 3 = _____   8 x 5 = _____

L. 6 x 6 = _____   8 x 4 = _____   3 x 7 = _____

M. 7 x 5 = _____   7 x 6 = _____   2 x 8 = _____

N. 7 x 7 = _____   8 x 9 = _____

O. 8 x 3 = _____   6 x 7 = _____

P. 9 x 9 = _____   8 x 6 = _____

Q. 9 x 6 = _____   9 x 5 = _____

© Frank Schaffer Publications, Inc.   FS-32071 Fourth Grade Math Review

Name _____   Multiplication facts

Tune-up

Write the products.

A.   5   7   8   6
     x 6   x 9   x 8   x 7

B.   1   5   6   1   4   3
     x 7   x 9   x 8   x 8   x 7   x 6

C.   2   8   4   3   8   5
     x 9   x 7   x 6   x 9   x 6   x 8

D.   3   2   3   8   7   5
     x 6   x 7   x 8   x 9   x 6   x 7

E.   7   6   7   1   9   2
     x 7   x 9   x 8   x 9   x 7   x 6

F.   3   6   4   9   9   9
     x 7   x 6   x 8   x 8   x 6   x 9

Name _____   Multiplication facts

 # Fact Finder

Circle groups of numbers that make multiplication facts. Look horizontally and vertically.

4	1	9	3	27	0	3	1	1	8	9
4	1	1	1	0	9	7	63	0	7	1
16	2	3	5	4	20	21	0	1	56	9
0	2	0	7	6	42	1	7	7	49	1
8	4	32	1	1	8	3	24	0	4	0
1	0	3	3	9	1	0	5	1	9	0
8	1	0	6	5	30	1	6	6	36	4
1	0	5	0	45	4	7	28	2	4	8
42	8	8	64	1	0	5	7	35	1	9
6	1	40	1	7	7	5	1	0	1	2
8	9	72	5	3	15	25	0	6	3	18

© Frank Schaffer Publications, Inc.   FS-32071 Fourth Grade Math Review

Name _____ Multiplication facts

Product Power

Complete the table.

X	1	7	5	9	8	2	6	0	3	4
6	6					12				
8		56							24	
		0		0		0		0		0
3										
		7		9		2		0	3	
7										28
9										
	5	25				10			15	20
4				36						
	2		10			4			6	8

© Frank Schaffer Publications, Inc.   32   FS-32071 Fourth Grade Math Review

Name _____ Mental multiplication

 **Special Products**

Use mental math to find the products.

A. 4 x 3 = _12_ 4 x 30 = _120_ 4 x 300 = _1,200_	B. 5 x 2 = _____ 5 x 20 = _____ 5 x 200 = _____
C. 7 x 1 = _____ 7 x 10 = _____ 7 x 100 = _____	D. 3 x 6 = _____ 3 x 60 = _____ 3 x 600 = _____
E. 2 x 9 = _____ 2 x 90 = _____ 2 x 900 = _____	F. 8 x 5 = _____ 8 x 50 = _____ 8 x 500 = _____
G. 6 x 4 = _____ 6 x 40 = _____ 6 x 400 = _____	H. 9 x 7 = _____ 9 x 70 = _____ 9 x 700 = _____

I.   20   50   60   40   70   60
      x 3   x 5   x 2   x 4   x 5   x 7

J.   40   80   20   50   30   90
      x 7   x 3   x 8   x 6   x 5   x 4

© Frank Schaffer Publications, Inc.   FS-32071 Fourth Grade Math Review

Name_____ Mental multiplication

 **Patterned Products**

Use mental math to find the products.

A.	3 x 3 — 9	30 x 3 — 90	300 x 3 — 900	B.	4 x 2	40 x 2	400 x 2
C.	4 x 3	40 x 3	400 x 3	D.	5 x 5	50 x 5	500 x 5
E.	6 x 2	60 x 2	600 x 2	F.	9 x 3	90 x 3	900 x 3
G.	8 x 4	80 x 4	800 x 4	H.	7 x 3	70 x 3	700 x 3
I.	2 x 7	20 x 7	200 x 7	J.	6 x 6	60 x 6	600 x 6

K. 20   400   100   60   500
   x 3   x 2   x 5   x 3   x 5

L. Write a sentence telling about the pattern you found.

_____

Name _____   Estimating products

Balanced Budgets

Round each number to its highest place. (Do not round one-digit numbers.) For each problem, multiply the rounded numbers to find the estimated product.

A. 7 x 43 =
   7 x 40 = 280
   _____

   5 x 88 =
   _____

B. 4 x 67 =
   _____

   6 x 12 =
   _____

C. 9 x 581 =
   _____

   3 x 297 =
   _____

D. 5 x 42 =
   _____

   8 x 26 =
   _____

E. 2 x 317 =
   _____

   5 x 697 =
   _____

Estimate the product.

F.   38   53   69   784   520
     x 4   x 8   x 5   x 3   x 7
     160

G.   81   59   26   679   809
     x 6   x 8   x 2   x 3   x 9

© Frank Schaffer Publications, Inc.   FS-32071 Fourth Grade Math Review

Name_____ Estimating products

Crystal Ball

Round each number to its highest place. (Do not round one-digit numbers.) For each problem, multiply the rounded numbers to find the estimated product.

A. 4 x 29 =
    4 x 30 = 120

3 x 87 =

B. 2 x 46 =

7 x 52 =

C. 3 x 279 =

6 x 819 =

D. 4 x 31 =

5 x 39 =

E. 3 x 888 =

8 x 312 =

Estimate the product.

F. 91   28   37   49   61
   x 6   x 7   x 2   x 5   x 8
   540

G. 385   421   868   217   775
   x 5   x 6   x 4   x 3   x 8

H. 269   496   340
   x 9   x 8   x 7

© Frank Schaffer Publications, Inc.   36   FS-32071 Fourth Grade Math Review

Name_____ Multiplying 2 or 3 digits by 1 digit

What's the Story?

Find the products. Regroup when you need to.

A. ²45   48   13   56   42
   x 4   x 3   x 9   x 3   x 2
   ---   ---   ---   ---   ---
   180

B.   75   58   75   14   23
   x 3   x 5   x 6   x 8   x 7
   ---   ---   ---   ---   ---

C.   29   81   36   57   15
   x 4   x 6   x 4   x 7   x 7
   ---   ---   ---   ---   ---

D. $ 0.84 $ 0.63 $ 0.92 $ 0.38 $ 0.21
   x   8   x   5   x   4   x   6   x   9
   ------   ------   ------   ------   ------

E. $ 0.47 $ 0.54 $ 0.92 $ 0.37 $ 0.32
   x   4   x   4   x   6   x   5   x   6
   ------   ------   ------   ------   ------

F. Write a story problem for 3 x 75. Then find the answer.

_____

_____

_____

© Frank Schaffer Publications, Inc.   37   FS-32071 Fourth Grade Math Review

Name _____   Multiplying 2 or 3 digits by 1 digit

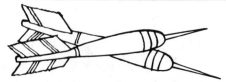

Target Practice

Multiply. Regroup when you need to.
Circle the products that would round to 200 or $2.00.

A.   96   23   52   43   75
      x 3   x 3   x 3   x 6   x 4
     288

B.   36   64   25   86   18
      x 4   x 5   x 7   x 3   x 6

C.   45   58   72   94   31
      x 6   x 4   x 8   x 7   x 6

D.   89   68   47   29   62
      x 9   x 2   x 3   x 4   x 5

E. $ 0.78 $ 0.67 $ 0.28 $ 0.43 $ 0.83
      x 5   x 4   x 7   x 9   x 3

F. $ 0.93 $ 0.53 $ 0.84 $ 0.27 $ 0.43
      x 3   x 8   x 6   x 6   x 8

© Frank Schaffer Publications, Inc.   FS-32071 Fourth Grade Math Review

Name _____ Multiplying 3 digits by 1 digit

Join the Parade

Multiply.

A. 164   412   172   198   958
   x   2   x   3   x   4   x   2   x   4
   ─────
    328

B. 426   583   395   108   282
   x   5   x   6   x   3   x   9   x   3

C. 501   362   614   492   354
   x   6   x   2   x   3   x   2   x   4

D. 347   461   927   519   863
   x   5   x   6   x   3   x   7   x   8

E. $ 2.14 $ 5.07 $ 1.61 $ 2.83 $ 6.95
   x   3   x   3   x   5   x   8   x   6

F. $ 1.26 $ 3.49 $ 6.12 $ 3.16 $ 2.08
   x   9   x   5   x   4   x   3   x   3

Name _____   Multiplying 3 digits by 1 digit

 ## It's What's Inside That Counts

Multiply. Then solve the riddle.

1. ²315 x 4 1,260 A	206 x 9 D	344 x 7 F	746 x 4 J	376 x 2 P
2. 941 x 5 G	503 x 7 M	626 x 5 W	328 x 7 O	642 x 3 H
3. 708 x 5 R	569 x 3 L	121 x 8 T	936 x 7 V	861 x 4 S
4. 473 x 2 B	613 x 4 Y	748 x 9 N	712 x 6 C	271 x 7 E

What 8-letter word has only 1 letter in it?

 A   ___   ___   ___   ___   ___   ___   ___
1,260 6,732 1,897 6,732 6,552 1,897 1,707 2,296 752 1,897

© Frank Schaffer Publications, Inc.   40   FS-32071 Fourth Grade Math Review

Name _____   Multiplying 4 digits by 1 digit

Picture-Perfect Products

Find each product.

A.
   1,413
x   4
―――――
   5,652

   7,738
x   2

   6,184
x   5

B.
   2,023
x   6

   5,670
x   3

   2,994
x   8

   6,092
x   3

C.
   1,232
x   4

   5,033
x   9

   2,943
x   5

   3,438
x   3

D.
   2,886
x   8

   4,433
x   6

   1,709
x   4

   6,330
x   2

E.
  $ 46.06
x   5

  $ 21.01
x   4

  $ 51.23
x   7

  $ 68.93
x   3

F.
  $ 41.05
x   6

  $ 73.14
x   8

  $ 43.28
x   4

  $ 68.43
x   7

© Frank Schaffer Publications, Inc.   FS-32071 Fourth Grade Math Review

Name _____   Multiplying 4 digits by 1 digit

Production Line

Find each product.

A. 4,792   3,412   3,017   2,999
   x   4   x   2   x   6   x   4
   19,168

B. 2,419   1,372   4,953   7,621
   x   2   x   4   x   2   x   8

C. 7,420   7,231   2,095   2,645
   x   6   x   6   x   8   x   3

D. 2,018   1,343   1,917   5,473
   x   5   x   2   x   4   x   5

E. $ 23.45 $ 64.92 $ 19.95 $ 23.06
   x   9 x   7 x   5 x   6

F. $ 75.55 $ 19.99
   x   3 x   6

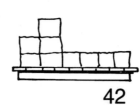

42

Name _____ Division facts

Quotient Quiz

Divide.

A. $18 \div 3 =$ __6__ $10 \div 2 =$ _____ $21 \div 3 =$ _____

B. $15 \div 5 =$ _____ $12 \div 3 =$ _____ $8 \div 2 =$ _____

C. $4 \div 2 =$ _____ $7 \div 1 =$ _____ $9 \div 3 =$ _____

D. $10 \div 5 =$ _____ $6 \div 2 =$ _____ $15 \div 3 =$ _____

E. $24 \div 4 =$ _____ $18 \div 2 =$ _____ $30 \div 5 =$ _____

F. $4 \div 4 =$ _____ $16 \div 4 =$ _____ $8 \div 4 =$ _____

G. $45 \div 5 =$ _____ $12 \div 2 =$ _____ $3 \div 3 =$ _____

H. $5 \div 1 =$ _____ $16 \div 2 =$ _____ $36 \div 4 =$ _____

I. $1\overline{)6}$ $2\overline{)10}$ $3\overline{)6}$ $4\overline{)12}$ $5\overline{)5}$

J. $3\overline{)3}$ $5\overline{)40}$ $4\overline{)32}$ $2\overline{)18}$ $1\overline{)1}$

K. $4\overline{)28}$ $5\overline{)35}$ $3\overline{)18}$

L. $5\overline{)25}$ $4\overline{)20}$ $4\overline{)24}$

© Frank Schaffer Publications, Inc.   43   FS-32071 Fourth Grade Math Review

Name _____   Division facts

 ## Driving Division

Write the quotients.

A. 4)2̄4̄ (6)   2)6̄   3)1̄2̄   4)4̄   1)5̄

B. 3)1̄8̄   2)1̄6̄   3)3̄   1)6̄   5)1̄0̄

C. 4)8̄   3)6̄   5)1̄5̄   3)2̄4̄   3)9̄

D. 2)4̄   5)5̄   2)8̄   1)1̄   4)1̄2̄

E. 12 ÷ 3 = _____   16 ÷ 4 = _____   36 ÷ 4 = _____

F. 18 ÷ 2 = _____   8 ÷ 1 = _____   10 ÷ 2 = _____

G. 35 ÷ 5 = _____   14 ÷ 2 = _____   25 ÷ 5 = _____

H. 15 ÷ 3 = _____   32 ÷ 4 = _____   7 ÷ 7 = _____

I. 24 ÷ 4 = _____   20 ÷ 5 = _____   28 ÷ 4 = _____

J. 45 ÷ 5 = _____   12 ÷ 2 = _____   21 ÷ 3 = _____

K. 20 ÷ 4 = _____   30 ÷ 5 = _____   4 ÷ 1 = _____

L. 27 ÷ 3 = _____   2 ÷ 2 = _____   40 ÷ 5 = _____

© Frank Schaffer Publications, Inc.   FS-32071 Fourth Grade Math Review

Name _____   Division facts

Pattern Paths

Write the quotients for each row. Look for a pattern.
Then write a division fact that continues the pattern.

A. 6)12   6)18   6)24   6)30   6)36 (quotient 6)

B. 5)30   6)36   7)42   8)48   )

C. 7)63   7)56   7)49   7)42   )

D. 9)81   8)72   7)63   6)54   )

E. 3)24   4)32   5)40   6)48   )

F. 9)63   8)56   7)49   6)42   )

G. 8)72   8)64   8)56   8)48   )

H. 5)15   6)18   7)21   8)24   )

© Frank Schaffer Publications, Inc.   45   FS-32071 Fourth Grade Math Review

Name _____ Division facts

Marble Mania

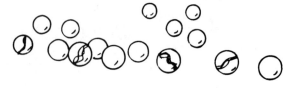

Write each quotient.

A. 18 ÷ 6 = __3__   27 ÷ 9 = _____   40 ÷ 8 = _____

B. 14 ÷ 7 = _____   8 ÷ 8 = _____   30 ÷ 6 = _____

C. 36 ÷ 6 = _____   6 ÷ 6 = _____   32 ÷ 8 = _____

D. 28 ÷ 7 = _____   12 ÷ 6 = _____   7 ÷ 7 = _____

E. 42 ÷ 7 = _____   21 ÷ 7 = _____   45 ÷ 9 = _____

F.   9 ÷ 9 = _____   36 ÷ 9 = _____   24 ÷ 8 = _____

G. 24 ÷ 6 = _____   54 ÷ 9 = _____   35 ÷ 7 = _____

H. 16 ÷ 8 = _____   18 ÷ 9 = _____   48 ÷ 8 = _____

I. 9)63   6)48   7)56   9)81   9)54

J. 8)64   7)42   6)42   7)63   8)72

K. 6)54   7)49   8)56   9)72   7)35

© Frank Schaffer Publications, Inc.   46   FS-32071 Fourth Grade Math Review

Name _____   Division facts

A Rule to Live By

Divide.

1.	T 9)9	A 6)30	T 2)2	
2.	F 9)81	R 8)56	E 9)18	H 7)42
3.	A 7)35	F 3)27	R 9)63	C 9)36
4.	E 8)16	K 6)48	E 7)14	O 8)24
5.	H 6)36	A 9)45	K 5)40	O 3)9
6.	F 6)54	R 7)49	A 8)40	H 8)48

Write the letters on the lines to find a good rule to follow.

__ __ __ __   __ __ __ __   __ __
1   5 8 2   4 5 7 2   3 9

__ __ __   __ __ __ __ __
1 6 2   2 5 7 1 6

47

© Frank Schaffer Publications, Inc.   FS-32071 Fourth Grade Math Review

Name _____ Division facts

Wishful Thinking

Write each quotient.
Shade the space if the quotient is 5 or less.

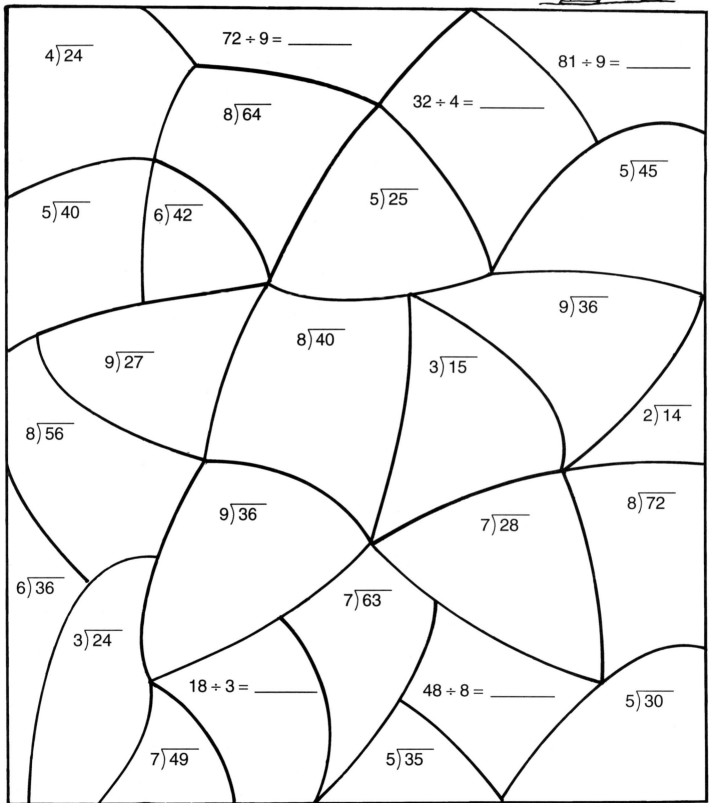

48

© Frank Schaffer Publications, Inc.   FS-32071 Fourth Grade Math Review

Name _____   Fact families

All of a Kind Family

Write four equations for each fact family.

A. 4, 8, 32
$4 \times 8 = 32$
$8 \times 4 = 32$
$32 \div 4 = 8$
$32 \div 8 = 4$

B. 3, 7, 21

C. 5, 8, 40

D. 6, 7, 42

E. 5, 9, 45

F. 6, 8, 48

G. 8, 9, 72

H. 9, 7, 63

Name _____ Fact families

Family Portrait

Multiply and divide.
Write the three numbers that make up each fact family.

A. 3 x 6 = ___18___
    6 x 3 = ___18___
   18 ÷ 3 = ___6___
   18 ÷ 6 = ___3___

3	6	18

B. 7 x 4 = _____
    4 x 7 = _____
   28 ÷ 7 = _____
   28 ÷ 4 = _____

C. 5 x 9 = _____
    9 x 5 = _____
   45 ÷ 5 = _____
   45 ÷ 9 = _____

D. 8 x 6 = _____
    6 x 8 = _____
   48 ÷ 8 = _____
   48 ÷ 6 = _____

E. 7 x 6 = _____
    6 x 7 = _____
   42 ÷ 7 = _____
   42 ÷ 6 = _____

F. 8 x 9 = _____
    9 x 8 = _____
   72 ÷ 8 = _____
   72 ÷ 9 = _____

G. 9 x 7 = _____
    7 x 9 = _____
   63 ÷ 9 = _____
   63 ÷ 7 = _____

H. 8 x 4 = _____
    4 x 8 = _____
   32 ÷ 8 = _____
   32 ÷ 4 = _____

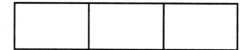

© Frank Schaffer Publications, Inc.   FS-32071 Fourth Grade Math Review

Quotient Patterns

Use mental math to find the quotients.

A. 15 ÷ 3 = __5__   150 ÷ 3 = __50__   1,500 ÷ 3 = __500__

B. 8 ÷ 2 = _____   80 ÷ 2 = _____   800 ÷ 2 = _____

C. 12 ÷ 4 = _____   120 ÷ 4 = _____   1,200 ÷ 4 = _____

D. 20 ÷ 5 = _____   200 ÷ 5 = _____   2,000 ÷ 5 = _____

E. 24 ÷ 6 = _____   240 ÷ 6 = _____   2,400 ÷ 6 = _____

F. 18 ÷ 3 = _____   180 ÷ 3 = _____   1,800 ÷ 3 = _____

G. 21 ÷ 7 = _____   210 ÷ 7 = _____   2,100 ÷ 7 = _____

H. 30 ÷ 6 = _____   300 ÷ 6 = _____   3,000 ÷ 6 = _____

I. 16 ÷ 4 = _____   160 ÷ 4 = _____   1,600 ÷ 4 = _____

J. 42 ÷ 7 = _____   420 ÷ 7 = _____   4,200 ÷ 7 = _____

K. 36 ÷ 6 = _____   360 ÷ 6 = _____   3,600 ÷ 6 = _____

L. 28 ÷ 4 = _____   280 ÷ 4 = _____   2,800 ÷ 4 = _____

Find each quotient.

M. 2)60   4)400   9)180   5)250   6)1,800

N. 4)800   8)240   5)1,000   2)120   8)1,600

Name _____ Special quotients

Special Quotients

Use mental math to find the quotients.

A. 12 ÷ 3 = __4__ 120 ÷ 3 = __40__ 1,200 ÷ 3 = __400__	B. 8 ÷ 4 = _____ 80 ÷ 4 = _____ 800 ÷ 4 = _____
C. 15 ÷ 5 = _____ 150 ÷ 5 = _____ 1,500 ÷ 5 = _____	D. 9 ÷ 3 = _____ 90 ÷ 3 = _____ 900 ÷ 3 = _____
E. 24 ÷ 4 = _____ 240 ÷ 4 = _____ 2,400 ÷ 4 = _____	F. 35 ÷ 7 = _____ 350 ÷ 7 = _____ 3,500 ÷ 7 = _____
G. 20 ÷ 5 = _____ 200 ÷ 5 = _____ 2,000 ÷ 5 = _____	H. 48 ÷ 6 = _____ 480 ÷ 6 = _____ 4,800 ÷ 6 = _____

I.   3)60   5)100   8)240   7)140   6)1,200

J.   4)800   3)150   2)100   5)2,500   9)2,700

© Frank Schaffer Publications, Inc.   FS-32071 Fourth Grade Math Review

Name _____   Estimating quotients

Friendly Numbers

Use compatible numbers to estimate each quotient. Compatible numbers are combinations of numbers that are easy to divide mentally.

A. 322 ÷ 4 =

322 is close to 320. Since 32 ÷ 4 = 8, 320 ÷ 4 = 80

  320 ÷ 4 = 80

B. 123 ÷ 3 =
_____

147 ÷ 5 =
_____

C. $2.41 ÷ 6 =
_____

$6.31 ÷ 7 =
_____

D. 117 ÷ 4 =
_____

267 ÷ 9 =
_____

E. $3.63 ÷ 9 =
_____

$4.51 ÷ 5 =
_____

F. 103 ÷ 5 =
_____

237 ÷ 8 =
_____

G. $4.19 ÷ 6 =
_____

$5.43 ÷ 6 =
_____

H. 477 ÷ 8 =
_____

143 ÷ 2 =
_____

I. $8.07 ÷ 9 =
_____

183 ÷ 3 =
_____

© Frank Schaffer Publications, Inc.   53   FS-32071 Fourth Grade Math Review

Name _____   Estimating quotients

Over or Under

Use compatible numbers to estimate each quotient. Compatible numbers are combinations of numbers that are easy to divide mentally.

628 is close to 630. Since 63 ÷ 9 = 7, 630 ÷ 9 = 70

A. 628 ÷ 9 =
630 ÷ 9 = 70

B. 318 ÷ 8 =

C. 244 ÷ 3 =

D. $8.96 ÷ 3 =

E. 149 ÷ 5 =

F. $3.96 ÷ 8 =

G. 643 ÷ 8 =

H. 358 ÷ 9 =

I. $5.61 ÷ 7 =

83 ÷ 4 =

477 ÷ 8 =

209 ÷ 7 =

$4.19 ÷ 7 =

723 ÷ 9 =

$8.11 ÷ 9 =

492 ÷ 7 =

324 ÷ 4 =

$5.39 ÷ 9 =

© Frank Schaffer Publications, Inc.   54   FS-32071 Fourth Grade Math Review

Name _____ Dividing 2 digits by 1 digit

Snow Fun

Divide. Each quotient has a remainder.

A. 4)22 with 5 R 2 shown, −20, 2   4)26   6)50   2)13

B. 3)14   5)23   4)19   7)17

C. 4)31   7)23   3)19   6)51

D. 6)37   8)67   9)15   5)39

E. 7)64   9)59

© Frank Schaffer Publications, Inc.   55   FS-32071 Fourth Grade Math Review

Name _____   Dividing 2 digits by 1 digit

Picnic Planning

Divide. Each quotient has a remainder.

A. 5)22̄ 4 R 2 / −20 / 2   6)39̄   5)33̄   8)76̄

B. 2)17̄   3)13̄   4)25̄   9)30̄

C. 8)65̄   2)11̄   6)27̄   7)53̄

D. 6)34̄   7)12̄   8)52̄   9)68̄

E. 7)59̄   4)26̄   9)44̄   8)31̄

© Frank Schaffer Publications, Inc.   56   FS-32071 Fourth Grade Math Review

Name _____ Dividing 2 digits by 1 digit

Butterfly Collector

Find the quotients.
Be sure to write the remainders.

A.
```
      17 R1
   4)69
     -4
     ‾‾
     29
     28
     ‾‾
      1
```
6)80   5)97

B. 3)49   7)78   8)99   6)82

C. 4)86   3)94   2)59   4)87

D. 3)71   4)73   5)88   7)16

Name _____ Dividing 2 digits by 1 digit

Hold On!

Find the quotients.
Be sure to write the remainders.

A.
```
    12 R1
 4)49
   -4
   ─────
    09
   - 8
   ─────
     1
```
3)67   5)73

B. 2)39   3)76   4)53   5)98

C. 3)95   2)49   6)86   7)79

D. 4)63   5)93   8)97   6)82

© Frank Schaffer Publications, Inc.   FS-32071 Fourth Grade Math Review

Name _____   Dividing 3 or 4 digits by 1 digit

Round and Round

Divide. Then use the code to solve the riddle.

Ⓐ
1. 9)128 → 14 R 2
   -9

   38
   -36

   2

Ⓝ 3)$3.33

Ⓚ 9)487

Ⓜ 6)723

Ⓤ
2. 8)762

Ⓢ 7)4,280

Ⓡ 3)$14.40

Ⓣ 9)643

Ⓟ
3. 8)659

Ⓔ 5)$15.95

Where would you find the world's biggest wheel?

A   A
--- ---   --- ---
14 R2 71 R4   14 R2 $1.11

A
--- --- --- --- --- --- --- ---
14 R2 120 R3 95 R2 611 R3 $3.19 120 R3 $3.19 $1.11 71 R4

                    A
     --- --- --- ---
     82 R3 14 R2 $4.80 54 R1

© Frank Schaffer Publications, Inc.   59   FS-32071 Fourth Grade Math Review

Name _____ Dividing 3 or 4 digits by 1 digit

Volcanic Ash Remains

Divide.

A.

$$6)\overline{769} \quad \begin{array}{r} 128 \text{ R}1 \\ -6 \\ \hline 16 \\ -12 \\ \hline 49 \\ -48 \\ \hline 1 \end{array}$$

9)461   5)746   4)3,926

B.

8)5,032   9)8,441   6)8,257   5)4,654

C.

5)758   8)1,712   6)550   9)8,572

D.

7)5,894   4)2,716   6)4,536   3)9,256

Name _____ Dividing 3 or 4 digits by 1 digit

Mystery Bag

Find the quotients.

A. 3)‾123   8)‾911   4)‾875   5)‾763

B. 8)‾$2.00   4)‾205   4)‾987   5)‾$20.50

C. 4)‾$64.04   8)‾369   2)‾751   5)‾127

D. 4)‾435   7)‾$0.84   9)‾832   6)‾$7.26

© Frank Schaffer Publications, Inc.   61   FS-32071 Fourth Grade Math Review

Name _____   Dividing 3 or 4 digits by 1 digit

Sticker Shares

Divide.

A. 5)632   4)948   8)825   6)$3.24

B. 7)806   3)552   9)987   8)$8.88

C. 3)1,891   5)129   6)609   6)101

D. 2)$16.10   9)$7.02   7)639   4)2,523

Name _____   Finding averages

Odd and Even Tug of War

Find the average for each group of numbers. Circle the averages.

A. 12, 35, 34, 15
 	
  12   ㉔
  35 4)96
  34   -8
        --- 	
  +15   16
  ---   -16
  96   ---
         0
	B. 39, 44, 84, 33	C. 121, 116, 132

D. 36, 10, 33, 45	E. 4, 5, 4, 9, 8	F. 214, 376, 148
G. 21, 36, 14, 13	H. 137, 275, 215	I. 62, 41, 77

How many averages were odd? _____

How many averages were even? _____

Which team won the tug of war? _____

© Frank Schaffer Publications, Inc.   63   FS-32071 Fourth Grade Math Review

Name _____ Finding averages

Find the Lucky Number

Find the average for each group of numbers. Then shade the triangle with the matching number. The unshaded triangle contains the lucky number!

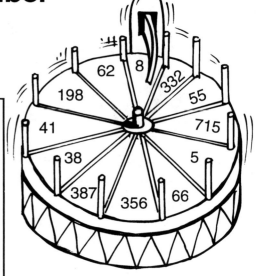

A. 64, 72, 99, 13 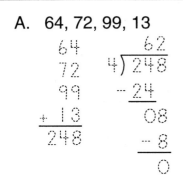	B. 5, 9, 7, 16, 3	
C. 241, 123, 632	D. 22, 44, 66, 88	E. 522, 811, 812
F. 2, 5, 7, 4, 7	G. 68, 83, 47	H. 387, 451, 323
I. 17, 38, 59	J. 24, 40, 59	K. 293, 279, 111, 109

© Frank Schaffer Publications, Inc.   64   FS-32071 Fourth Grade Math Review

Name _____ Understanding decimal concepts

A Shady Story

Shade each diagram as indicated. Then write a fraction and a decimal to show what part of the diagram is shaded.

A. four tenths

   $\frac{4}{10}$

   0.4

B. five tenths

   _____

   _____

C. one and three tenths

    _____

   _____

D. eight tenths

   _____

   _____

E. one and nine tenths

    _____

   _____

F. fifty-four hundredths

   _____

   _____

G. one and eleven hundredths

    _____

   _____

H. ninety-two hundredths

   _____

   _____

I. one and eighty-one hundredths

    _____

   _____

J. seventeen hundredths

   _____

   _____

K. one and four hundredths

    _____

   _____

© Frank Schaffer Publications, Inc.   FS-32071 Fourth Grade Math Review

Name _____ Understanding decimal concepts

What's the Point?

Write a fraction and a decimal for each number or diagram.

A. two tenths $\frac{2}{10}$ 0.2   one and nine tenths _____ _____

B. six tenths _____ _____   two and four tenths _____ _____

C. five tenths _____ _____   three and one tenth _____ _____

D. fifty-seven hundredths _____ _____

E. seventy-two hundredths _____ _____

F. ninety-one hundredths _____ _____

G. one and eighty-six hundredths _____ _____

H. six and fourteen hundredths _____ _____

I. nine and three hundredths _____ _____

J.   K.   L.

_____   _____   _____

M. 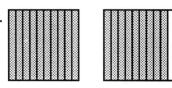 N.   O.

_____   _____   _____

P.   Q.   R.

_____   _____   _____

Name_____   Comparing and ordering decimals

Down the Garden Path

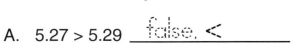

Read each statement and write **true** or **false**. If the statement is false, write the symbol that would make it true.

A. 5.27 > 5.29 _false, <_   0.09 = 0.90 _____

B. 8.2 = 8.20 _____   17 < 16.9 _____

C. 0.80 = 0.8 _____   7.16 > 7.2 _____

D. 0.23 < 0.32 _____   6.05 < 6.50 _____

E. 2.91 < 2.8 _____   3 < 3.01 _____

F. 6.95 < 7.02 _____   0.66 > 0.6 _____

G. 8.71 > 8.8 _____   9.89 > 10 _____

H. 2.89 = 2.9 _____   6.19 > 6.2 _____

I. 4 = 4.00 _____   3.78 < 3.79 _____

Begin at Start. Shade the boxes to show a path in order from the least number to the greatest. You may move left, right, up, or down.

0.05	0.08	0.07	0.14	0.73	1.2	1.23	1.29
0.01	0.1	0.09	0.94	1.0	1.06	1.05	1.33
0.19	0.12	0.02	0.8	0.17	0.21	0.97	1.41
0.25	0.11	0.03	0.63	0.16	0.15	0.62	1.56
0.3	0.32	0.43	0.51	0.2	0.16	0.99	2.0

© Frank Schaffer Publications, Inc.   FS-32071 Fourth Grade Math Review

Name _____   Comparing and ordering decimals

May I Take Your Order, Please?

Write >, < or = in the ◯ to complete each number sentence.

A.	6.23	◯	6.4	15.4	◯	16	9.29	◯	9.2
B.	0.4	◯	0.40	1.2	◯	1.19	3.7	◯	3.8
C.	1.98	◯	2	3	◯	3.0	8.8	◯	8.88
D.	0.38	◯	0.29	4.6	◯	4.59	6.0	◯	6.00
E.	8.91	◯	8.9	2.36	◯	2.4	7.20	◯	7.2
F.	6.44	◯	7.0	5.97	◯	6.01	10.0	◯	9.99
G.	1.23	◯	1.32	15.4	◯	1.54	8.93	◯	8.9

Arrange each group of numbers to make six decimal numbers.
Arrange the decimal numbers in order from the least to the greatest.

H. 0, 5, 9 _____ _____ _____ _____ _____ _____

I. 1, 4, 7 _____ _____ _____ _____ _____ _____

J. 6, 3, 8 _____ _____ _____ _____ _____ _____

K. 2, 4, 9 _____ _____ _____ _____ _____ _____

L. 5, 8, 0 _____ _____ _____ _____ _____ _____

M. 3, 7, 1 _____ _____ _____ _____ _____ _____

N. 6, 0, 4 _____ _____ _____ _____ _____ _____

© Frank Schaffer Publications, Inc.   68   FS-32071 Fourth Grade Math Review

Name _____   Rounding decimals

Where's the Whole?

Round each decimal to the nearest whole number.

A. 3.4 _3_   4.8 _____   8.2 _____   3.7 _____

B. 0.8 _____   1.1 _____   9.5 _____   6.3 _____

C. 16.2 _____   25.3 _____   84.9 _____   93.6 _____

D. 29.4 _____   73.1 _____   19.8 _____   48.7 _____

E. 3.72 _____   4.88 _____   9.14 _____   5.63 _____

F. 6.97 _____   17.83 _____   26.26 _____   71.17 _____

G. 49.65 _____   99.51 _____   2.71 _____   16.61 _____

H. 1.47 _____   23.09 _____   88.88 _____   4.06 _____

I. 25.11 _____   2.55 _____   46.36 _____   39.10 _____

J. 99.91 _____   6.49 _____   52.50 _____   53.16 _____

Look at the decimal numbers in the box below. Write each decimal number beside the whole number that it would round to.

54.8	53.75	55.25	52.95
53.26	54.5	54.05	53.19
55.43	53.49	53.99	53.87

K. 53 _____ _____ _____ _____

L. 54 _____ _____ _____ _____

M. 55 _54.8_ _____ _____ _____

69

Name _____   Rounding decimals

The Whole Truth

Round each decimal number to the nearest whole number.

A.	2.9	3.2	0.6	6.7	8.1
	3				

B.	71.3	16.4	94.5	29.6	31.3

C.	81.6	43.5	58.2	46.7	96.8

D.	2.73	3.64	4.13	8.88	4.95

E.	16.82	27.28	99.51	19.62	36.49

F.	12.65	63.14	52.83	76.09	15.34

G.	80.92	27.77	55.68	29.83	10.44

Write three numbers that round to each given number.

H. 10 _____   _____   _____

I. 95 _____   _____   _____

© Frank Schaffer Publications, Inc.   70   FS-32071 Fourth Grade Math Review

Name _____   Adding and subtracting decimals

Ducky Decimals

Find the sums or differences. Watch the signs!

A.   3.4   45.3   8.4   40.8
     + 6.1   + 21.9   – 6.2   + 67.4
      9.5

B.   7.7   83.3   55.6   71.6
     – 4.6   – 24.5   + 47.5   – 55.5

C. $ 53.70   0.88   43.75 $ 0.67
   + 85.68   – 0.48   + 82.19   + 0.81

D. $ 8.04   0.76 $ 1.79   62.83
   – 3.26   – 0.59   + 1.79   – 29.77

E.   51.0 – 6.4   82.61 – 14.48 $63.00 – 7.27
     51.0
    – 6.4
     44.6

F.   48.06 + 3.41   4.09 – 0.79 $0.64 + $0.69

© Frank Schaffer Publications, Inc.   FS-32071 Fourth Grade Math Review

Name _____   Adding and subtracting decimals

Decimal Dots

Add or subtract. Watch the signs!

A.
    6.3
  + 4.5
    10.8

    7.8
  + 6.4

  15.45
+ 6.19

B.
$ 7.02
− 2.65

  64.2
+ 28.7

  17.6
+ 33.8

  0.72
− 0.59

C.
$ 57.20
− 4.84

  0.89
+ 0.53

  6.02
− 0.69

  24.61
+ 16.18

D.
$ 35.09
− 2.73

$ 72.14
+ 28.35

$ 36.05
− 9.18

  60.47
+ 35.64

E. 8.5 + 2.7
    8.5
  + 2.7
    11.2

15.3 − 9.6

$5.31 − $0.82

F. 30.54 − 8.38

5.46 + 12.77

$35.19 + 23.81

Name _____ Equivalent fractions

Feed the Birds

Write the missing numbers to make each pair of fractions equivalent.

A.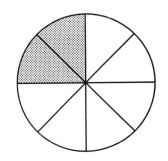

$\frac{1}{4} = \frac{2}{8}$

B.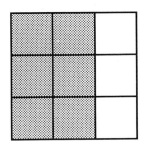

$\frac{2}{3} = \frac{}{9}$

C.

$\frac{3}{4} = \frac{}{16}$

D.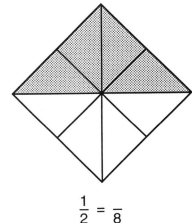

$\frac{1}{2} = \frac{}{8}$

E.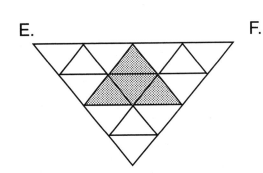

$\frac{1}{4} = \frac{}{16}$

F.

$\frac{1}{5} = \frac{}{15}$

Multiply each numerator and denominator by 2, 3, and 4 to find three equivalent fractions.

G. $\frac{2}{3} = \frac{4}{6} = \frac{}{} = \frac{}{}$

H. $\frac{2}{5} = \frac{}{} = \frac{}{} = \frac{}{}$

I. $\frac{1}{4} = \frac{}{} = \frac{}{} = \frac{}{}$

J. $\frac{3}{8} = \frac{}{} = \frac{}{} = \frac{}{}$

K. $\frac{4}{5} = \frac{}{} = \frac{}{} = \frac{}{}$

© Frank Schaffer Publications, Inc.   FS-32071 Fourth Grade Math Review

Name _____ Equivalent fractions

Shades, Anyone?

Shade the circle at the right to show a fraction that is equivalent to the circle at the left. Then write the equivalent fraction.

A.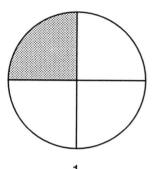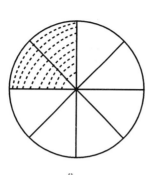

$\dfrac{1}{4} = \dfrac{2}{8}$

B.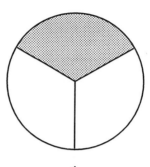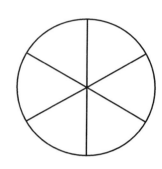

$\dfrac{1}{3} = \dfrac{}{6}$

C.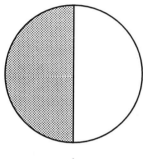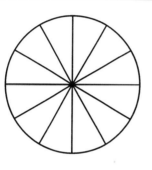

$\dfrac{1}{2} = \dfrac{}{12}$

D.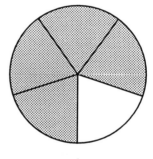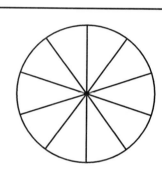

$\dfrac{4}{5} = \dfrac{}{10}$

E.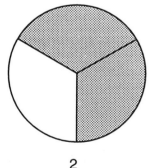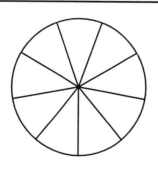

$\dfrac{2}{3} = \dfrac{}{9}$

F.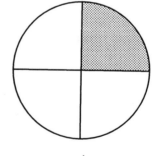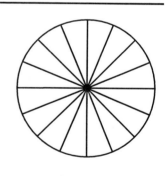

$\dfrac{1}{4} = \dfrac{}{16}$

Multiply each numerator and denominator by 2, 3, and 4 to find three equivalent fractions.

G. $\dfrac{1}{2} = \dfrac{2}{4} = \dfrac{}{} = \dfrac{}{}$

H. $\dfrac{5}{8} = \dfrac{}{} = \dfrac{}{} = \dfrac{}{}$

I. $\dfrac{1}{5} = \dfrac{}{} = \dfrac{}{} = \dfrac{}{}$

J. $\dfrac{3}{4} = \dfrac{}{} = \dfrac{}{} = \dfrac{}{}$

© Frank Schaffer Publications, Inc.   FS-32071 Fourth Grade Math Review

Name _____   Simplest form fractions

 # Limbo Lowdown

Write each fraction in its simplest form. A fraction is in its simplest form when the only number that will divide both the numerator and the denominator is 1.

A. $\frac{3}{12} = \frac{1}{4}$ $\frac{6}{18} = $ ___ $\frac{9}{27} = $ ___

B. $\frac{4}{6} = $ ___ $\frac{8}{10} = $ ___ $\frac{12}{14} = $ ___ $\frac{2}{8} = $ ___

C. $\frac{18}{27} = $ ___ $\frac{18}{24} = $ ___ $\frac{16}{30} = $ ___ $\frac{9}{21} = $ ___

D. $\frac{21}{28} = $ ___ $\frac{27}{63} = $ ___ $\frac{15}{25} = $ ___ $\frac{18}{20} = $ ___

Shade each box that contains a fraction in its simplest form.

E.

$\frac{18}{20}$	$\frac{4}{6}$	$\frac{6}{18}$	$\frac{5}{8}$	$\frac{2}{3}$	$\frac{5}{15}$	$\frac{3}{9}$	$\frac{11}{12}$	$\frac{7}{9}$	$\frac{14}{18}$	$\frac{18}{22}$	$\frac{9}{36}$
$\frac{5}{9}$	$\frac{10}{40}$	$\frac{6}{12}$	$\frac{4}{5}$	$\frac{7}{10}$	$\frac{4}{12}$	$\frac{21}{28}$	$\frac{22}{25}$	$\frac{5}{12}$	$\frac{3}{6}$	$\frac{16}{24}$	$\frac{9}{16}$
$\frac{10}{12}$	$\frac{3}{8}$	$\frac{45}{50}$	$\frac{10}{15}$	$\frac{9}{15}$	$\frac{8}{12}$	$\frac{7}{14}$	$\frac{20}{30}$	$\frac{30}{36}$	$\frac{8}{16}$	$\frac{8}{13}$	$\frac{4}{8}$
$\frac{14}{16}$	$\frac{25}{30}$	$\frac{4}{7}$	$\frac{7}{8}$	$\frac{18}{27}$	$\frac{10}{20}$	$\frac{6}{10}$	$\frac{20}{40}$	$\frac{5}{6}$	$\frac{2}{5}$	$\frac{4}{12}$	$\frac{28}{30}$
$\frac{27}{30}$	$\frac{16}{18}$	$\frac{5}{15}$	$\frac{21}{24}$	$\frac{9}{10}$	$\frac{2}{7}$	$\frac{9}{13}$	$\frac{8}{11}$	$\frac{18}{20}$	$\frac{3}{12}$	$\frac{15}{18}$	$\frac{6}{30}$

© Frank Schaffer Publications, Inc.   FS-32071 Fourth Grade Math Review

Name _____ Simplest form fractions

Zoo Fun

Rewrite each fraction in its simplest form. Then use the code letters to solve the riddle.

A $\frac{4}{6} = \frac{2}{3}$	E $\frac{10}{12} = $ ___	I $\frac{20}{25} = $ ___	N $\frac{4}{12} = $ ___
J $\frac{7}{28} = $ ___	O $\frac{8}{56} = $ ___	U $\frac{27}{36} = $ ___	G $\frac{9}{24} = $ ___
M $\frac{16}{32} = $ ___	L $\frac{16}{36} = $ ___	H $\frac{14}{24} = $ ___	S $\frac{15}{25} = $ ___
T $\frac{20}{28} = $ ___	W $\frac{63}{81} = $ ___	K $\frac{50}{80} = $ ___	

What do you get when you cross a hyena with a parrot?

$\frac{A}{\frac{2}{3}}$ $\frac{}{\frac{1}{3}}$ $\frac{A}{\frac{2}{3}}$ $\frac{}{\frac{1}{3}}$ $\frac{}{\frac{4}{5}}$ $\frac{}{\frac{1}{2}}$ $\frac{A}{\frac{2}{3}}$ $\frac{}{\frac{4}{9}}$

$\frac{}{\frac{5}{7}}$ $\frac{}{\frac{7}{12}}$ $\frac{A}{\frac{2}{3}}$ $\frac{}{\frac{5}{7}}$ $\frac{}{\frac{4}{9}}$ $\frac{A}{\frac{2}{3}}$ $\frac{}{\frac{3}{4}}$ $\frac{}{\frac{3}{8}}$ $\frac{}{\frac{7}{12}}$ $\frac{}{\frac{3}{5}}$

$\frac{A}{\frac{2}{3}}$ $\frac{}{\frac{5}{7}}$ $\frac{}{\frac{4}{5}}$ $\frac{}{\frac{5}{7}}$ $\frac{}{\frac{3}{5}}$ $\frac{}{\frac{1}{7}}$ $\frac{}{\frac{7}{9}}$ $\frac{}{\frac{1}{3}}$

$\frac{}{\frac{1}{4}}$ $\frac{}{\frac{1}{7}}$ $\frac{}{\frac{5}{8}}$ $\frac{}{\frac{5}{6}}$ $\frac{}{\frac{3}{5}}$

Name _____ Comparing and ordering fractions

Size It Up

Compare the fractions in each pair. To do this, change one or both fractions so that they have the same denominator. Then write <, >, or = in each ◯.

A. $\frac{3}{8}$ ◯ $\frac{1}{2}$ $\frac{1}{2}$ ◯ $\frac{1}{4}$ $\frac{3}{9}$ ◯ $\frac{6}{18}$

B. $\frac{2}{3}$ ◯ $\frac{4}{5}$ $\frac{1}{4}$ ◯ $\frac{2}{8}$ $\frac{3}{4}$ ◯ $\frac{5}{12}$

C. $\frac{1}{5}$ ◯ $\frac{1}{4}$ $\frac{5}{8}$ ◯ $\frac{2}{3}$ $\frac{3}{8}$ ◯ $\frac{1}{4}$

D. $\frac{1}{10}$ ◯ $\frac{1}{5}$ $\frac{3}{4}$ ◯ $\frac{2}{3}$ $\frac{1}{3}$ ◯ $\frac{4}{6}$

E. $\frac{2}{3}$ ◯ $\frac{3}{8}$ $\frac{2}{10}$ ◯ $\frac{1}{5}$ $\frac{2}{9}$ ◯ $\frac{3}{4}$

F. $\frac{1}{2}$ ◯ $\frac{1}{3}$ $\frac{3}{10}$ ◯ $\frac{1}{4}$ $\frac{2}{5}$ ◯ $\frac{3}{10}$

G. $\frac{2}{5}$ ◯ $\frac{3}{4}$ $\frac{3}{4}$ ◯ $\frac{5}{8}$ $\frac{1}{4}$ ◯ $\frac{1}{10}$

H. $\frac{3}{10}$ ◯ $\frac{1}{5}$ $\frac{3}{8}$ ◯ $\frac{2}{5}$ $\frac{3}{5}$ ◯ $\frac{3}{4}$

I. $\frac{2}{5}$ ◯ $\frac{4}{10}$ $\frac{5}{8}$ ◯ $\frac{3}{10}$ $\frac{5}{8}$ ◯ $\frac{3}{5}$

Write the fractions in order on each number line.

J. $\frac{1}{4}$, $\frac{1}{8}$, $\frac{3}{8}$, $\frac{3}{4}$, $\frac{7}{8}$

K. $\frac{5}{12}$, $\frac{11}{12}$, $\frac{5}{6}$, $\frac{1}{6}$, $\frac{1}{12}$

Name _____   Comparing and ordering fractions

Lead the Way

Compare the fractions in each pair. To do this, change one or both fractions so that they have the same denominator. Then write <, >, or = in each ◯.

A. $\frac{1}{5}$ ◯ $\frac{1}{4}$ $\frac{5}{6}$ ◯ $\frac{9}{10}$

B. $\frac{3}{5}$ ◯ $\frac{4}{5}$ $\frac{5}{7}$ ◯ $\frac{9}{10}$ $\frac{2}{4}$ ◯ $\frac{1}{2}$

C. $\frac{2}{5}$ ◯ $\frac{1}{2}$ $\frac{7}{8}$ ◯ $\frac{3}{4}$ $\frac{3}{10}$ ◯ $\frac{1}{4}$

D. $\frac{3}{4}$ ◯ $\frac{2}{3}$ $\frac{9}{12}$ ◯ $\frac{3}{4}$ $\frac{8}{10}$ ◯ $\frac{4}{5}$

E. $\frac{9}{10}$ ◯ $\frac{7}{8}$ $\frac{5}{7}$ ◯ $\frac{3}{4}$ $\frac{2}{3}$ ◯ $\frac{3}{5}$

F. $\frac{1}{3}$ ◯ $\frac{2}{3}$ $\frac{2}{5}$ ◯ $\frac{2}{3}$ $\frac{8}{12}$ ◯ $\frac{2}{3}$

G. $\frac{1}{3}$ ◯ $\frac{1}{2}$ $\frac{2}{4}$ ◯ $\frac{4}{8}$ $\frac{9}{16}$ ◯ $\frac{3}{8}$

In each box below, draw arrows in order from the least fraction to the greatest fraction.

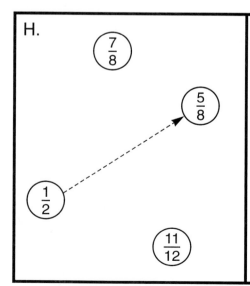

H. $\frac{7}{8}$, $\frac{5}{8}$, $\frac{1}{2}$, $\frac{11}{12}$

I. $\frac{1}{6}$, $\frac{1}{2}$, $\frac{1}{3}$, $\frac{1}{4}$

J. $\frac{2}{5}$, $\frac{2}{3}$, $\frac{1}{4}$, $\frac{7}{10}$, $\frac{19}{20}$

© Frank Schaffer Publications, Inc.   FS-32071 Fourth Grade Math Review

Name _____ Finding a fraction of a number

Take Part

Find the fractional part of each number.

A.

$\frac{1}{2}$ of 10 = __5__ $\frac{3}{4}$ of 8 = _____

Draw a diagram to find the fractional part of each number.

B. $\frac{3}{5}$ of 20 = _____ $\frac{1}{3}$ of 21 = _____ $\frac{5}{6}$ of 12 = _____

C. $\frac{1}{7}$ of 28 = _____ $\frac{2}{3}$ of 6 = _____ $\frac{1}{2}$ of 18 = _____

D. $\frac{3}{4}$ of 12 = _____ $\frac{1}{5}$ of 20 = _____ $\frac{6}{7}$ of 14 = _____

E. $\frac{1}{8}$ of 16 = _____ $\frac{2}{3}$ of 12 = _____ $\frac{3}{4}$ of 16 = _____

F. $\frac{1}{4}$ of 12 = _____ $\frac{1}{2}$ of 20 = _____ $\frac{1}{10}$ of 10 = _____

Name _____ Finding a fraction of a number

Parts of the Pie

Find the fraction of the number.

A.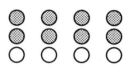

$\frac{2}{3}$ of 12 = __8__ $\frac{3}{5}$ of 10 = _____

Draw a diagram to find the fractional part of each number.

B. $\frac{1}{3}$ of 18 = _____ $\frac{1}{2}$ of 8 = _____ $\frac{2}{3}$ of 15 = _____

C. $\frac{1}{2}$ of 10 = _____ $\frac{3}{8}$ of 8 = _____ $\frac{1}{4}$ of 20 = _____

D. $\frac{9}{10}$ of 20 = _____ $\frac{1}{4}$ of 16 = _____ $\frac{4}{7}$ of 14 = _____

E. $\frac{2}{3}$ of 21 = _____ $\frac{3}{8}$ of 16 = _____ $\frac{4}{5}$ of 10 = _____

F. $\frac{5}{8}$ of 16 = _____ $\frac{3}{5}$ of 20 = _____ $\frac{3}{4}$ of 16 = _____

© Frank Schaffer Publications, Inc.   FS-32071 Fourth Grade Math Review

Name _____ Improper fractions and mixed numbers

All Mixed-up

Draw diagrams to help you rewrite each improper fraction as a mixed number.

A. $\frac{7}{4} = 1\frac{3}{4}$ $\frac{4}{3} = $ _____ $\frac{9}{4} = $ _____

B. $\frac{8}{3} = $ _____ $\frac{10}{3} = $ _____ $\frac{11}{4} = $ _____ $\frac{9}{2} = $ _____

C. $\frac{9}{3} = $ _____ $\frac{14}{3} = $ _____ $\frac{15}{4} = $ _____ $\frac{12}{3} = $ _____

Draw diagrams to help you rewrite each mixed number as an improper fraction.

D. $1\frac{3}{5} = $ _____ $3\frac{1}{2} = $ _____ $2\frac{1}{3} = $ _____ $2\frac{3}{4} = $ _____

E. $3\frac{3}{4} = $ _____ $3\frac{2}{3} = $ _____ $1\frac{7}{8} = $ _____ $4\frac{1}{2} = $ _____

© Frank Schaffer Publications, Inc.   81   FS-32071 Fourth Grade Math Review

Name _____ Improper fractions and mixed numbers

Trail Mix

Draw an arrow from each mixed number to its matching diagram and improper fraction.

	Mixed Number		Improper Fraction
A.	$3\frac{1}{4}$		$\frac{5}{2}$
B.	$2\frac{1}{2}$		$\frac{10}{3}$
C.	$4\frac{2}{5}$		$\frac{13}{4}$
D.	$3\frac{1}{3}$		$\frac{23}{8}$
E.	$2\frac{7}{8}$		$\frac{22}{5}$
F.	$4\frac{2}{3}$		$\frac{25}{6}$
G.	$5\frac{3}{4}$		$\frac{14}{3}$
H.	$4\frac{1}{6}$		$\frac{16}{5}$
I.	$3\frac{1}{5}$		$\frac{23}{4}$
J.	$1\frac{5}{12}$		$\frac{17}{12}$
K.	$2\frac{3}{5}$		$\frac{24}{5}$
L.	$4\frac{4}{5}$		$\frac{13}{5}$

Name _____   Relating fractions and decimals

Changing Forms

Write a decimal for each fraction.

A. $\frac{2}{4}$ = __0.5__ $\frac{1}{5}$ = _____ $\frac{3}{10}$ = _____

B. $\frac{4}{5}$ = _____ $\frac{6}{10}$ = _____ $\frac{6}{25}$ = _____

C. $\frac{12}{50}$ = _____ $\frac{8}{10}$ = _____ $\frac{4}{20}$ = _____

D. $\frac{67}{100}$ = _____ $\frac{13}{50}$ = _____ $\frac{3}{4}$ = _____

Write a fraction in simplest form for each decimal.

E. 0.5 = _____   0.23 = _____   0.4 = _____

F. 0.8 = _____   0.25 = _____

G. 0.2 = _____   0.88 = _____

H. 0.75 = _____   0.28 = _____

Write a fraction and a decimal for each picture.

I.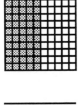

   _____   _____   _____

   _____   _____   _____

Name _____   Relating fractions and decimals

Part Partners

Write a decimal for each fraction.

A. $\frac{1}{2}$ = 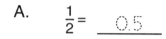 $\frac{3}{4}$ = _____

B. $\frac{5}{10}$ = _____ $\frac{15}{25}$ = _____ $\frac{6}{20}$ = _____

C. $\frac{2}{5}$ = _____ $\frac{12}{20}$ = _____ $\frac{1}{10}$ = _____

D. $\frac{2}{25}$ = _____ $\frac{13}{20}$ = _____ $\frac{4}{5}$ = _____

E. $\frac{9}{10}$ = _____ $\frac{47}{100}$ = _____ $\frac{9}{50}$ = _____

Write a fraction in simplest form for each decimal.

F. 0.6 = $\frac{6}{10} = \frac{3}{5}$ 0.24 = _____   0.43 = _____

G. 0.4 = _____   0.75 = _____   0.50 = _____

H. 0.65 = _____   0.30 = _____   0.33 = _____

I. 0.9 = _____   0.16 = _____   0.25 = _____

Shade the diagrams to show the fractional or decimal parts.

J.   0.15   K. $\frac{1}{2}$ L. $\frac{2}{5}$

© Frank Schaffer Publications, Inc.   84   FS-32071 Fourth Grade Math Review

Name_____   Adding fractions

Mysterious Mix

To add fractions with like denominators, add the numerators. Write each sum in simplest form. Shade the matching section on the design below.

A. $\dfrac{5}{9} + \dfrac{2}{9} = $ _____ $\dfrac{3}{8} + \dfrac{4}{8} = $ _____

B. $\dfrac{1}{8} + \dfrac{4}{8} = $ _____ $\dfrac{1}{10} + \dfrac{6}{10} = $ _____ $\dfrac{3}{5} + \dfrac{1}{5} = $ _____

C. $\dfrac{1}{7} + \dfrac{3}{7} = $ _____ $\dfrac{3}{12} + \dfrac{7}{12} = $ _____ $\dfrac{1}{4} + \dfrac{2}{4} = $ _____

D. $\dfrac{4}{9} + \dfrac{2}{9} = $ _____ $\dfrac{1}{12} + \dfrac{5}{12} = $ _____ $\dfrac{2}{8} + \dfrac{1}{8} = $ _____

E. $\dfrac{2}{5} + \dfrac{1}{5} = $ _____ $\dfrac{3}{13} + \dfrac{4}{13} = $ _____ $\dfrac{4}{15} + \dfrac{1}{15} = $ _____

F. $3\dfrac{1}{3}$ $2\dfrac{2}{5}$ $5\dfrac{3}{4}$ $8\dfrac{1}{7}$
  $+1\dfrac{1}{3}$ $+3\dfrac{1}{5}$ $+2$ $+1\dfrac{2}{7}$

G. $6\dfrac{3}{8}$ $4\dfrac{2}{9}$
  $+1\dfrac{1}{8}$ $+3\dfrac{1}{9}$

H. $2\dfrac{3}{10}$ $1\dfrac{3}{6}$
  $+3\dfrac{2}{10}$ $+2\dfrac{2}{6}$

I. $3\dfrac{4}{12}$ $2\dfrac{3}{10}$
  $+1\dfrac{5}{12}$ $+\dfrac{2}{10}$

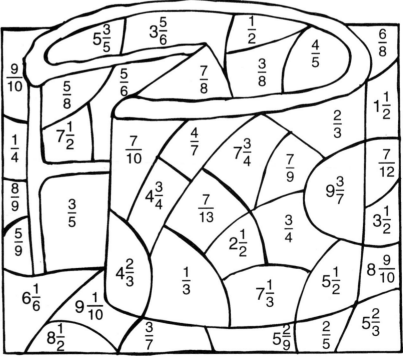

85

Name_____ Adding fractions

Stormy Weather

To add fractions with like denominators, add the numerators. Write each sum in simplest form. Then use the code to solve the riddle.

1. $\frac{1}{4} + \frac{2}{4} =$ _____ $\frac{1}{5} + \frac{3}{5} =$ _____ $\frac{7}{9} + \frac{1}{9} =$ _____
              A   B   C

2. $\frac{3}{10} + \frac{2}{10} =$ _____ $\frac{2}{7} + \frac{3}{7} =$ _____ $\frac{1}{10} + \frac{5}{10} =$ _____
              D   E   F

3. $\frac{2}{9} + \frac{4}{9} =$ _____ $\frac{2}{5} + \frac{1}{5} =$ _____ $\frac{2}{12} + \frac{6}{12} =$ _____
              G   H   I

4. $\frac{1}{12} + \frac{3}{12} =$ _____ $\frac{1}{15} + \frac{4}{15} =$ _____ $\frac{3}{10} + \frac{6}{10} =$ _____
              J   K   L

5. $3\frac{1}{4}$ $2\frac{1}{4}$ $3\frac{1}{6}$ $5\frac{3}{5}$
     $+1\frac{2}{4}$ $+3\frac{2}{4}$ $+2\frac{2}{6}$ $+2\frac{1}{5}$
        M   N   O   P

6. $2\frac{1}{8}$ $3\frac{1}{8}$ $1\frac{4}{9}$ $4\frac{2}{10}$
     $+2\frac{1}{8}$ $+3$ $+3\frac{1}{9}$ $+3\frac{4}{10}$
        R   S   T   U

What goes up when the rain comes down?

___ ___
$\frac{3}{4}$ $5\frac{3}{4}$

___ ___ ___ ___ ___ ___ ___ ___
$7\frac{3}{5}$ $4\frac{3}{4}$ $\frac{4}{5}$ $4\frac{1}{4}$ $\frac{5}{7}$ $\frac{9}{10}$ $\frac{9}{10}$ $\frac{3}{4}$

© Frank Schaffer Publications, Inc.   FS-32071 Fourth Grade Math Review

Name _____ Subtracting fractions

Action Fractions

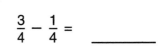

Write each difference in its simplest form.

A. $\dfrac{7}{8} - \dfrac{3}{8} = \dfrac{4}{8} = \dfrac{1}{2}$ $\dfrac{5}{7} - \dfrac{4}{7} =$ _____

B. $\dfrac{6}{10} - \dfrac{3}{10} =$ _____ $\dfrac{2}{3} - \dfrac{1}{3} =$ _____ $\dfrac{3}{4} - \dfrac{1}{4} =$ _____

C. $\dfrac{4}{5} - \dfrac{2}{5} =$ _____ $\dfrac{5}{6} - \dfrac{2}{6} =$ _____ $\dfrac{9}{10} - \dfrac{8}{10} =$ _____

D. $\dfrac{7}{8} - \dfrac{1}{8} =$ _____ $\dfrac{8}{9} - \dfrac{6}{9} =$ _____ $\dfrac{11}{15} - \dfrac{1}{15} =$ _____

E. $\dfrac{3}{5} - \dfrac{2}{5} =$ _____ $\dfrac{9}{10} - \dfrac{4}{10} =$ _____ $\dfrac{9}{12} - \dfrac{4}{12} =$ _____

F. $\dfrac{6}{10} - \dfrac{4}{10} =$ _____ $\dfrac{2}{9} - \dfrac{1}{9} =$ _____ $\dfrac{8}{12} - \dfrac{2}{12} =$ _____

G. $\begin{array}{r} 3\frac{2}{4} \\ -1\frac{1}{4} \\ \hline 2\frac{1}{4} \end{array}$ $\begin{array}{r} 5\frac{3}{4} \\ -2\frac{1}{4} \\ \hline \end{array}$ $\begin{array}{r} 4\frac{9}{10} \\ -3\frac{1}{10} \\ \hline \end{array}$ $\begin{array}{r} 6\frac{7}{9} \\ -4 \\ \hline \end{array}$

H. $\begin{array}{r} 9\frac{3}{5} \\ -6\frac{1}{5} \\ \hline \end{array}$ $\begin{array}{r} 2\frac{5}{6} \\ -1\frac{3}{6} \\ \hline \end{array}$ $\begin{array}{r} 5\frac{4}{7} \\ -3 \\ \hline \end{array}$ $\begin{array}{r} 6\frac{7}{10} \\ -5\frac{2}{10} \\ \hline \end{array}$

I. $\begin{array}{r} 4\frac{7}{8} \\ -\frac{3}{8} \\ \hline \end{array}$ $\begin{array}{r} 6\frac{6}{8} \\ -3\frac{3}{8} \\ \hline \end{array}$ $\begin{array}{r} 6\frac{11}{12} \\ -4\frac{2}{12} \\ \hline \end{array}$ $\begin{array}{r} 5\frac{7}{10} \\ -\frac{2}{10} \\ \hline \end{array}$

© Frank Schaffer Publications, Inc.   FS-32071 Fourth Grade Math Review

Name_____ Subtracting fractions

What's Missing?

Write each difference in its simplest form.

A. $\dfrac{7}{10} - \dfrac{5}{10} = \dfrac{2}{10} = \dfrac{1}{5}$ $\dfrac{2}{4} - \dfrac{1}{4} = $ _____

B. $\dfrac{4}{7} - \dfrac{2}{7} = $ _____ $\dfrac{4}{10} - \dfrac{2}{10} = $ _____ $\dfrac{9}{12} - \dfrac{1}{12} = $ _____

C. $\dfrac{7}{8} - \dfrac{3}{8} = $ _____ $\dfrac{12}{15} - \dfrac{7}{15} = $ _____ $\dfrac{5}{6} - \dfrac{2}{6} = $ _____

D. $\dfrac{2}{3} - \dfrac{1}{3} = $ _____ $\dfrac{5}{10} - \dfrac{1}{10} = $ _____ $\dfrac{8}{12} - \dfrac{2}{12} = $ _____

E. $\dfrac{5}{6} - \dfrac{3}{6} = $ _____ $\dfrac{4}{7} - \dfrac{1}{7} = $ _____ $\dfrac{3}{8} - \dfrac{1}{8} = $ _____

F. $\dfrac{4}{9} - \dfrac{1}{9} = $ _____ $\dfrac{7}{11} - \dfrac{3}{11} = $ _____ $\dfrac{4}{10} - \dfrac{3}{10} = $ _____

G. $\;\;3\dfrac{5}{8}\;\;\;\;\;\;\;\;\;\;\;\;\;\;6\dfrac{2}{5}\;\;\;\;\;\;\;\;\;\;\;\;\;\;9\dfrac{7}{10}\;\;\;\;\;\;\;\;\;\;\;\;\;\;4\dfrac{3}{7}$
$\;\;\;-1\dfrac{4}{8}\;\;\;\;\;\;\;\;\;\;\;-5\dfrac{1}{5}\;\;\;\;\;\;\;\;\;\;\;-3\dfrac{2}{10}\;\;\;\;\;\;\;\;\;\;\;-\dfrac{1}{7}$

H. $\;\;6\dfrac{7}{8}\;\;\;\;\;\;\;\;\;\;\;\;\;\;2\dfrac{4}{5}\;\;\;\;\;\;\;\;\;\;\;\;\;\;5\dfrac{7}{8}\;\;\;\;\;\;\;\;\;\;\;\;\;\;6\dfrac{1}{2}$
$\;\;\;-\dfrac{3}{8}\;\;\;\;\;\;\;\;\;\;\;\;-1\dfrac{3}{5}\;\;\;\;\;\;\;\;\;\;\;\;-\dfrac{5}{8}\;\;\;\;\;\;\;\;\;\;\;\;-5$

I. $\;\;4\dfrac{8}{10}\;\;\;\;\;\;\;\;\;\;\;\;\;5\dfrac{3}{4}\;\;\;\;\;\;\;\;\;\;\;\;\;\;8\dfrac{7}{9}\;\;\;\;\;\;\;\;\;\;\;\;\;\;9\dfrac{12}{15}$
$\;\;\;-3\dfrac{3}{10}\;\;\;\;\;\;\;\;\;\;-1\;\;\;\;\;\;\;\;\;\;\;\;\;\;\;\;-3\dfrac{1}{9}\;\;\;\;\;\;\;\;\;\;\;-3\dfrac{2}{15}$

© Frank Schaffer Publications, Inc.   FS-32071 Fourth Grade Math Review

Name _____   Adding fractions

Piece It Together

Change the fractions in each problem so that they have the same denominator. Then add the numerators and write the sum in its simplest form.

A. $\dfrac{1}{8} = \dfrac{1}{8}$ $\dfrac{1}{3}$ $\dfrac{7}{10}$
    $+\dfrac{3}{4} = +\dfrac{6}{8}$ $+\dfrac{1}{6}$ $+\dfrac{1}{5}$
    _____   _____   _____
         $\dfrac{7}{8}$

B. $\dfrac{1}{8}$ $\dfrac{1}{2}$ $\dfrac{3}{8}$
    $+\dfrac{1}{2}$ $+\dfrac{1}{6}$ $+\dfrac{1}{4}$
    _____   _____   _____

C. $\dfrac{3}{10}$ $\dfrac{1}{6}$ $\dfrac{1}{10}$
    $+\dfrac{2}{5}$ $+\dfrac{2}{3}$ $+\dfrac{3}{5}$
    _____   _____   _____

D. $\dfrac{1}{12}$ $\dfrac{4}{9}$ $\dfrac{1}{2}$
    $+\dfrac{1}{6}$ $+\dfrac{1}{3}$ $+\dfrac{1}{4}$
    _____   _____   _____

E. $1\dfrac{1}{3}$ $2\dfrac{1}{4}$ $3\dfrac{1}{8}$
    $+\ \dfrac{1}{6}$ $+1\dfrac{1}{8}$ $+2\dfrac{3}{16}$
    _____   _____   _____

F. $4\dfrac{1}{4}$ $3\dfrac{1}{10}$ $4\dfrac{2}{3}$
    $+1\dfrac{5}{12}$ $+5\dfrac{3}{5}$ $+2\dfrac{1}{12}$
    _____   _____   _____

© Frank Schaffer Publications, Inc.   FS-32071 Fourth Grade Math Review

Name _____   Adding fractions

   **Spare Parts**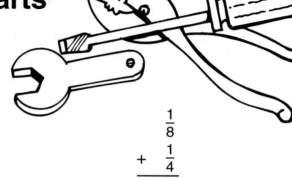

Change the fractions in each problem so that they have the same denominator. Then add the numerators and write the sum in its simplest form.

A. $\frac{5}{8} = \frac{5}{8}$ $\frac{1}{2}$ $\frac{1}{8}$
   $+\frac{1}{4} = +\frac{2}{8}$ $+\frac{3}{8}$ $+\frac{1}{4}$
   ─────────────   ─────────   ─────────
   $\frac{7}{8}$

B. $\frac{1}{3}$ $\frac{1}{2}$ $\frac{1}{2}$
   $+\frac{1}{6}$ $+\frac{1}{6}$ $+\frac{1}{8}$
   ─────────   ─────────   ─────────

C. $\frac{1}{3}$ $\frac{2}{3}$ $\frac{1}{4}$
   $+\frac{1}{9}$ $+\frac{1}{6}$ $+\frac{3}{8}$
   ─────────   ─────────   ─────────

D. $\frac{1}{2}$ $\frac{2}{5}$ $\frac{1}{3}$
   $+\frac{1}{10}$ $+\frac{3}{10}$ $+\frac{1}{12}$
   ─────────   ─────────   ─────────

E. $1\frac{1}{2}$ $4\frac{2}{3}$ $3\frac{1}{4}$
   $+1\frac{1}{4}$ $+2\frac{1}{6}$ $+1\frac{5}{12}$
   ─────────   ─────────   ─────────

F. $2\frac{1}{5}$ $3\frac{2}{3}$ $3\frac{3}{10}$
   $+\frac{2}{10}$ $+5\frac{1}{12}$ $+2\frac{1}{2}$
   ─────────   ─────────   ─────────

© Frank Schaffer Publications, Inc.   FS-32071 Fourth Grade Math Review

Name _____   Subtracting fractions

Cooking Calculations

Change the fractions in each problem so that they have the same denominator. Then subtract the numerators and write the difference in its simplest form.

A. $\dfrac{2}{3} = \dfrac{4}{6}$
$-\dfrac{1}{6} = -\dfrac{1}{6}$
$\overline{\dfrac{3}{6} = \dfrac{1}{2}}$

$\dfrac{7}{8}$
$-\dfrac{1}{4}$

$\dfrac{1}{2}$
$-\dfrac{1}{6}$

B. $\dfrac{5}{8}$
$-\dfrac{1}{2}$

$\dfrac{5}{9}$
$-\dfrac{1}{3}$

$\dfrac{9}{10}$
$-\dfrac{1}{2}$

C. $\dfrac{3}{4}$
$-\dfrac{3}{8}$

$\dfrac{1}{2}$
$-\dfrac{1}{4}$

$\dfrac{2}{3}$
$-\dfrac{2}{9}$

D. $\dfrac{4}{5}$
$-\dfrac{3}{10}$

$\dfrac{7}{8}$
$-\dfrac{3}{4}$

$\dfrac{1}{2}$
$-\dfrac{3}{8}$

E. $1\dfrac{1}{3}$
$-\dfrac{1}{6}$

$3\dfrac{9}{10}$
$-1\dfrac{1}{2}$

$5\dfrac{3}{8}$
$-2\dfrac{1}{4}$

F. $2\dfrac{3}{4}$
$-\dfrac{1}{2}$

$6\dfrac{7}{8}$
$-3\dfrac{3}{4}$

$7\dfrac{11}{12}$
$-5\dfrac{1}{6}$

© Frank Schaffer Publications, Inc.   91   FS-32071 Fourth Grade Math Review

Name _____   Subtracting fractions

Bits and Pieces

Change the fractions in each problem so that they have the same denominator. Then subtract the numerators and write the difference in its simplest form.

A. $\dfrac{5}{8} = \dfrac{5}{8}$
    $-\dfrac{1}{4} = -\dfrac{2}{8}$
    $\rule{2cm}{0.4pt}$
    $\dfrac{3}{8}$

    $\dfrac{1}{2}$
    $-\dfrac{3}{8}$

    $\dfrac{3}{4}$
    $-\dfrac{1}{8}$

B. $\dfrac{1}{3}$
    $-\dfrac{1}{6}$

    $\dfrac{3}{8}$
    $-\dfrac{1}{4}$

    $\dfrac{7}{10}$
    $-\dfrac{1}{5}$

C. $\dfrac{3}{4}$
    $-\dfrac{1}{2}$

    $\dfrac{5}{6}$
    $-\dfrac{1}{3}$

    $\dfrac{7}{8}$
    $-\dfrac{1}{2}$

D. $\dfrac{2}{3}$
    $-\dfrac{1}{6}$

    $\dfrac{8}{9}$
    $-\dfrac{1}{3}$

    $\dfrac{11}{12}$
    $-\dfrac{1}{2}$

E. $1\dfrac{1}{2}$
    $-\dfrac{1}{4}$

    $3\dfrac{3}{8}$
    $-1\dfrac{1}{4}$

    $5\dfrac{4}{5}$
    $-2\dfrac{1}{10}$

F. $6\dfrac{9}{10}$
    $-3\dfrac{2}{5}$

    $4\dfrac{2}{3}$
    $-\dfrac{1}{6}$

    $3\dfrac{7}{8}$
    $-1\dfrac{3}{4}$

© Frank Schaffer Publications, Inc.   92   FS-32071 Fourth Grade Math Review

Name _____   Multiplying 2-digit numbers

Ten-Gallon Products

Multiply.

A. 43   64   46   75   41
   x 20 x 10 x 50 x 80 x 30
   860

B. 16   89   53   76   36
   x 90 x 40 x 30 x 50 x 20

C. 95   35   67   52   83
   x 60 x 80 x 40 x 20 x 50

D. 26   41   76   66   39
   x 20 x 90 x 40 x 50 x 30

E. 46   83   63   76   55
   x 60 x 60 x 40 x 30 x 50

F. 18   84   92
   x 90 x 60 x 70

© Frank Schaffer Publications, Inc.   93   FS-32071 Fourth Grade Math Review

Name _____ Multiplying 2-digit numbers

Product Prizes

Find the products.

A. 36   51   41   17   85
   x 20   x 40   x 30   x 10   x 40
   720

B. 46   77   43   37   48
   x 30   x 90   x 70   x 70   x 60

C. 85   27   55   76   93
   x 50   x 90   x 80   x 40   x 70

D. 59   83   65   75   67
   x 50   x 80   x 90   x 50   x 80

E. 68   95   52   84   77
   x 30   x 80   x 20   x 70   x 30

F. 38   52   97
   x 40   x 60   x 50

© Frank Schaffer Publications, Inc.   94   FS-32071 Fourth Grade Math Review

Name _____ Multiplying 2-digit numbers

Environmental Products

Multiply. Use the products and the letters to find a message on the trash bin.

A 34 x 23 102 + 680 ――― 782	B 18 x 11	D 42 x 12	E 64 x 55
G 65 x 38	I 86 x 75	L 94 x 32	N 31 x 49
O 95 x 35	R 88 x 43	T 94 x 89	U 98 x 62

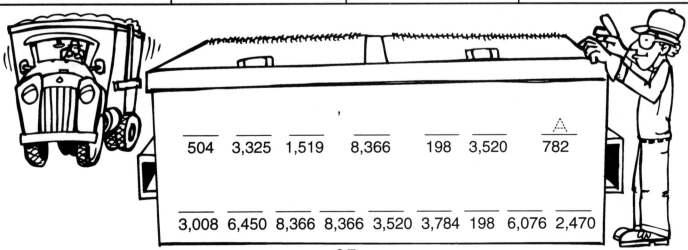

___ ___ ___ ___ ___ ___ ___A__
504 3,325 1,519 8,366 198 3,520 782

___ ___ ___ ___ ___ ___ ___ ___ ___
3,008 6,450 8,366 8,366 3,520 3,784 198 6,076 2,470

© Frank Schaffer Publications, Inc. 95 FS-32071 Fourth Grade Math Review

Name _____ Multiplying by 2-digit numbers

 ## Products in Paradise

Multiply.

A.
```
   1
  14        39        23        58
x 23      x 12      x 28      x 32
  42
+280
----
 322
```

B.
```
  44        13        45        46
x 76      x 29      x 63      x 23
```

C.
```
  36        59        32        77
x 37      x 14      x 65      x 22
```

D.
```
  48        64        33        28
x 32      x 23      x 19      x 17
```

© Frank Schaffer Publications, Inc.   96   FS-32071 Fourth Grade Math Review

Name _____   Multiplying by 2-digit numbers

Martian Multiplication

Find the products.

A. 83   92   51   16
   x 46   x 77   x 85   x 61
   498
  +3320
   ────
   3,818

B. 82   65   34   42
   x 76   x 54   x 24   x 67

C. 148   162   237   489
   x 53   x 53   x 19   x 33

D. 752   834   924   783
   x 25   x 68   x 26   x 79

E. 312   674
   x 58   x 85

97

© Frank Schaffer Publications, Inc.   FS-32071 Fourth Grade Math Review

Name _____   Multiplication

Product Puzzlers

Multiply. Use the products to complete the puzzle.

Across

A. 523
   x 64
   ‾‾‾‾
   2092
 +31380
 ‾‾‾‾‾‾
  33,472

F. 402
   x 12

J. 503
   x 72

K. 901
   x   8

L. 31
   x 2

M. 604
   x 38

P. 11
   x 6

Q. 53
   x 16

Down

A. 66
   x 51

B. 74
   x 49

C. 7
   x 6

D. 216
   x 33

E. 164
   x 16

F. 25
   x 19

G. 137
   x   6

H. 4
   x 5

I. 161
   x   3

N. 49
   x 2

98

© Frank Schaffer Publications, Inc.   FS-32071 Fourth Grade Math Review

Name _____   Dividing by 2-digit numbers

Upward Bound

Find the quotients and the remainders.

A. 6 R 13
   50)313
   -300
   ----
    13   40)84   80)249

B. 70)413   90)572   60)252   30)87

C. 40)556   20)793   30)678   50)961

D. 40)881   30)770   20)503   80)917

Name _____ Dividing by 2-digit numbers

Sailing Along

Find the quotients and the remainders.

A. 3 R 73
   90)343
   -270
   ‾‾‾
    73

50)369   40)95   80)179

B. 70)542   40)298   20)97   90)850

C. 60)683   10)131   30)415   20)593

D. 70)830   80)957   10)311

© Frank Schaffer Publications, Inc.   100   FS-32071 Fourth Grade Math Review

Name _____   Dividing by 2-digit numbers

"Whooo" Knows?

Find the quotients and the remainders.

A. 56)462 8 R 14
   −448
   ‾‾‾‾
   14

14)93

44)249

B. 36)267

41)352

26)79

53)245

C. 27)725

44)673

51)870

39)497

D. 32)680

17)473

65)800

13)277

Division Dinos

Find the quotients and the remainders.

A. 28)235̄ (8 R 11, −224, 11)   32)84̄   33)253̄   46)168̄

B. 19)44̄   31)78̄   63)525̄   58)426̄

C. 46)634̄   17)473̄   51)827̄   29)788̄

D. 34)538̄   37)892̄   28)793̄

Name _____ Multiplying and dividing with money

Balloon Bazaar

Multiply or divide to solve each problem.

A. $0.15 each
Find the cost of 47.
$0.15
x 47
105
600
$7.05

B. $0.28 each
Find the cost of 30.

C. $8.28 per dozen
Find the cost of 1.

D. $7.92 for 22
Find the cost of 1.

E. $5.04 per dozen
Find the cost of 1.

F. $1.89 each
Find the cost of 15.

G. $9.60 for 20
Find the cost of 1.

H. $2.85 for 15
Find the cost of 1.

I. $0.97 each
Find the cost of 42.

J. $2.29 each
Find the cost of 13.

© Frank Schaffer Publications, Inc. 103 FS-32071 Fourth Grade Math Review

Name _____   Multiplying and dividing with money

Farmers' Market

Use the information on the sign to solve the problems.

bananas	$0.69	pound	kiwi	$3.00	dozen
grapefruit	$0.17	pound	oranges	$5.55	15-pound bag
juice	$0.17	ounce	apples	$7.90	10-pound bag
cantaloupes	$1.19	each	grapes	$4.84	11-pound package
honeydew	$0.85	pound			

A. Find the cost of 15 pounds of bananas.

B. Find the cost of 20 pounds of grapefruit.

C. Find the cost of 1 kiwi.

D. Find the cost of 1 pound of oranges.

E. Find the cost of 1 pound of apples.

F. Find the cost of 24 ounces of juice.

G. Find the cost of 1 pound of grapes.

H. Find the cost of 14 cantaloupes.

I. Find the cost of 13 pounds of honeydew.

© Frank Schaffer Publications, Inc.   FS-32071 Fourth Grade Math Review

Answer Key

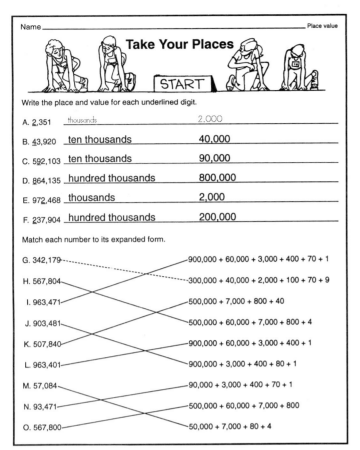

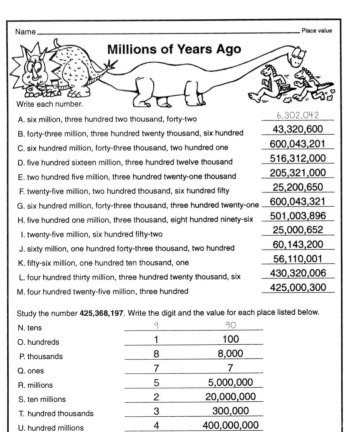

Page 1 · Page 2 · Page 3 · Page 4

105

© Frank Schaffer Publications, Inc.   FS-32071 Fourth Grade Math Review

Answer Key

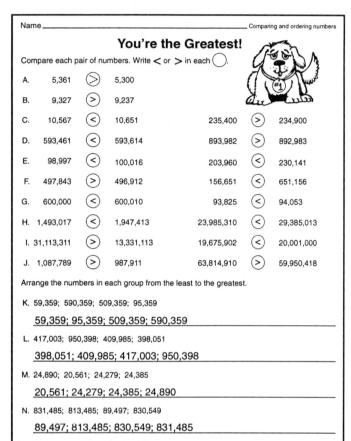

Answer Key

Fast-track Facts
Add or subtract.

A.	2 +9 = 11	6 +7 = 13	9 −8 = 1	14 −7 = 7	6 +5 = 11	7 +9 = 16
B.	16 −8 = 8	9 +4 = 13	7 +7 = 14	17 −9 = 8	11 −8 = 3	2 +9 = 11
C.	3 +9 = 12	15 −8 = 7	7 +3 = 10	4 +8 = 12	9 +8 = 17	10 −5 = 5
D.	13 −8 = 5	12 −9 = 3	4 +6 = 10	15 −6 = 9	12 −7 = 5	13 −6 = 7

E. 18 − 9 = 9   14 − 8 = 6   7 + 8 = 15
F. 14 − 5 = 9   6 + 9 = 15   5 + 7 = 12
G. 16 − 9 = 7   7 + 4 = 11   8 + 8 = 16
H. 17 − 8 = 9   9 + 5 = 14   8 + 6 = 14
I. 8 + 4 = 12   15 − 9 = 6
J. 16 − 7 = 9   5 + 8 = 13
K. 9 + 9 = 18   13 − 4 = 9

Page 9

Just the Facts
Find each sum or difference. Watch the signs!

A. 9 + 8 = 17   10 − 4 = 6   12 − 9 = 3
B. 16 − 8 = 8   5 + 8 = 13   6 + 6 = 12
C. 15 − 6 = 9   7 + 9 = 16   15 − 7 = 8
D. 12 − 6 = 6   9 + 2 = 11   8 + 6 = 14
E. 13 − 4 = 9   8 + 8 = 16   16 − 9 = 7
F. 6 + 7 = 13   10 − 2 = 8   12 − 3 = 9
G. 12 − 6 = 6   13 − 8 = 5   14 − 6 = 8

H.	4 +7 = 11	6 +4 = 10	11 −9 = 2	10 −9 = 1	7 +8 = 15	18 −9 = 9
I.	5 +5 = 10	14 −7 = 7	11 −8 = 3	12 −5 = 7	3 +9 = 12	2 +9 = 11
J.	9 +9 = 18	6 +8 = 14	10 −3 = 7	15 −9 = 6		
K.	14 −8 = 6	7 +8 = 15	5 +7 = 12	6 +9 = 15		

Page 10

In the Ballpark
Round each number to its highest place. Then estimate the sums and differences.

A. 61 (rounds to 60) + 73 (rounds to 70) = 130 (60 + 70 = 130)   92 −14 → 90 −10 = 80   41 +49 → 40 +50 = 90   60 −48 → 60 −50 = 10
B. 378 −103 → 400 −100 = 300   603 −485 → 600 −500 = 100   729 +196 → 700 +200 = 900   746 +718 → 700 +700 = 1,400
C. 927 +896 → 900 +900 = 1,800   842 −645 → 800 −600 = 200   903 −875 → 900 −900 = 0   296 +849 → 300 +800 = 1,100
D. 6,899 −2,463 → 7,000 −2,000 = 5,000   2,985 +3,109 → 3,000 +3,000 = 6,000   987 +4,106 → 1,000 +4,000 = 5,000   3,795 −1,128 → 4,000 −1,000 = 3,000
E. 7,537 +2,486 → 8,000 +2,000 = 10,000   9,205 −6,886 → 9,000 −7,000 = 2,000   8,831 +5,679 → 9,000 +6,000 = 15,000   7,015 −5,983 → 7,000 −6,000 = 1,000
F. 23,857 +19,452 → 20,000 +20,000 = 40,000   96,810 −28,563 → 100,000 −30,000 = 70,000   47,923 +13,526 → 50,000 +10,000 = 60,000   83,519 −30,274 → 80,000 −30,000 = 50,000

Page 11

In the Rough
Round each number to its highest place. Then estimate the sums.

A. 64 + 43 = (60 + 40 = 100) 100   78 + 52 = (80 + 50) 130
B. 51 + 36 = (50 + 40) 90   85 + 72 = (90 + 70) 160
C. 381 + 278 = (400 + 300) 700   144 + 585 = (100 + 600) 700
D. 562 + 897 = (600 + 900) 1,500   659 + 845 = (700 + 800) 1,500
E. 2,965 + 3,149 = (3,000 + 3,000) 6,000   1,782 + 3,946 = (2,000 + 4,000) 6,000
F. 6,854 + 7,310 = (7,000 + 7,000) 14,000   8,483 + 4,701 = (8,000 + 5,000) 13,000
G. 29,357 + 31,468 = (30,000 + 30,000) 60,000   65,826 + 12,419 = (70,000 + 10,000) 80,000
H. 43,925 + 28,046 = (40,000 + 30,000) 70,000   81,997 + 8,692 = (80,000 + 9,000) 89,000

Round each number to its highest place. Then estimate the differences.

I. 78 − 54 = (80 − 50) 30   53 − 29 = (50 − 30) 20
J. 69 − 61 = (70 − 60) 10   92 − 89 = (90 − 90) 0
K. 493 − 147 = (500 − 100) 400   525 − 237 = (500 − 200) 300
L. 263 − 98 = (300 − 100) 200   807 − 485 = (800 − 500) 300
M. 5,398 − 2,427 = (5,000 − 2,000) 3,000   9,872 − 5,188 = (10,000 − 5,000) 5,000
N. 8,312 − 7,903 = (8,000 − 8,000) 0   3,786 − 2,241 = (4,000 − 2,000) 2,000
O. 78,845 − 21,341 = (80,000 − 20,000) 60,000   94,378 − 52,481 = (90,000 − 50,000) 40,000
P. 56,012 − 39,085 = (60,000 − 40,000) 20,000   26,834 − 16,296 = (30,000 − 20,000) 10,000

Page 12

© Frank Schaffer Publications, Inc.   107   FS-32071 Fourth Grade Math Review

Answer Key

Page 13

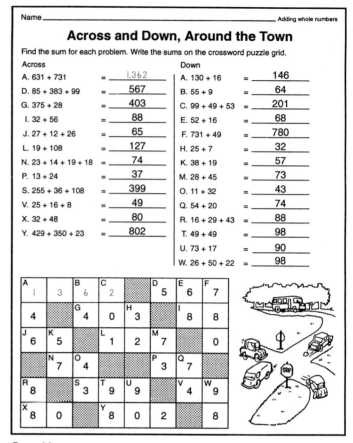

Page 14

Totally Cool

Find the sums.

A. 2,345 + 3,872 = 6,217   4,614 + 3,747 = 8,361   2,937 + 1,758 = 4,695   5,807 + 956 = 6,763

B. 6,821 + 59 = 6,880   2,436 + 3,827 = 6,263   2,385 + 3,561 = 5,946   5,661 + 3,359 = 9,020

C. 1,986 + 5,923 = 7,909   2,931 + 3,142 = 6,073   4,385 + 3,579 = 7,964   5,188 + 1,996 = 7,184

D. 352 + 7,863 = 8,215   2,481 + 6,879 = 9,360   3,105 + 4,896 = 8,001   4,385 + 2,961 = 7,346

E. $35.19 + $26.85 = $62.04 $68.75 + $9.97 = $78.72 $43.50 + $29.79 = $73.29 $38.46 + $17.38 = $55.84

F. 3,125 + 936 + 1,215 = 5,276   5,687 + 47 + 863 = 6,597 $23.95 + 8.97 + 15.60 = $48.52

Page 15

Cashier's Delight

Add.

A. 4,382 + 1,166 = 5,548   5,783 + 927 = 6,710   1,983 + 1,872 = 3,855   2,520 + 3,980 = 6,500

B. 7,193 + 1,875 = 9,068   1,735 + 3,961 = 5,696   2,961 + 1,446 = 4,407   896 + 8,473 = 9,369

C. 6,388 + 1,793 = 8,181   3,871 + 5,619 = 9,490   8,375 + 924 = 9,299   5,309 + 3,998 = 9,307

D. 4,986 + 1,194 = 6,180   9,758 + 162 = 9,920   8,234 + 1,398 = 9,632   3,685 + 4,839 = 8,524

E. $20.64 + $53.47 = $74.11 $59.41 + $38.64 = $98.05 $26.53 + $61.74 = $88.27 $89.95 + $9.97 = $99.92

F. 2,357 + 3,496 + 196 = 6,049   4,285 + 1,846 + 2,099 = 8,230 $15.65 + 8.27 + 42.93 = $66.85

Page 16

Answer Key

Page 17 — Larger-Than-Life (Adding whole numbers)

Find the sums.

A. 32,967 + 19,824 = 52,791 ; 63,937 + 4,852 = 68,789 ; 15,816 + 9,537 = 25,353 ; 89,471 + 3,895 = 93,366

B. 23,689 + 18,497 = 42,186 ; 25,963 + 17,827 = 43,790 ; 43,895 + 16,742 = 60,637 ; 56,140 + 19,396 = 75,536

C. 27,975 + 38,046 = 66,021 ; 56,910 + 9,438 = 66,348 ; 25,926 + 18,409 = 44,335 ; 18,888 + 47,196 = 66,084

D. $384.16 + $19.81 = $403.97 ; $109.83 + $120.79 = $230.62 ; $258.32 + $570.07 = $828.39 ; $585.46 + $18.54 = $604.00

E. 28,532 + 1,829 + 32,047 = 62,408 ; 58,896 + 1,746 + 25,814 = 86,456 ; 67,242 + 2,037 + 15,384 = 84,663

F. $199.36 + 270.46 + $89.95 = $559.77 ; $89.74 + 146.75 + $689.61 = $926.10 ; $356.89 + 28.34 + $563.57 = $948.80

Page 18 — The Great One (Adding whole numbers)

Add.

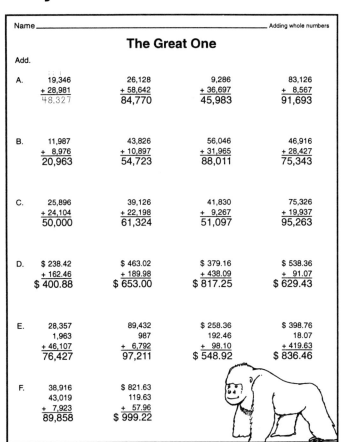

A. 19,346 + 28,981 = 48,327 ; 26,128 + 58,642 = 84,770 ; 9,286 + 36,697 = 45,983 ; 83,126 + 8,567 = 91,693

B. 11,987 + 8,976 = 20,963 ; 43,826 + 10,897 = 54,723 ; 56,046 + 31,965 = 88,011 ; 46,916 + 28,427 = 75,343

C. 25,896 + 24,104 = 50,000 ; 39,126 + 22,198 = 61,324 ; 41,830 + 9,267 = 51,097 ; 75,326 + 19,937 = 95,263

D. $238.42 + $162.46 = $400.88 ; $463.02 + $189.98 = $653.00 ; $379.16 + $438.09 = $817.25 ; $538.36 + $91.07 = $629.43

E. 28,357 + 1,963 + 46,107 = 76,427 ; 89,432 + 987 + 6,792 = 97,211 ; $258.36 + 192.46 + 98.10 = $548.92 ; $398.76 + 18.07 + 419.63 = $836.46

F. 38,916 + 43,019 + 7,923 = 89,858 ; $821.63 + 119.63 + 57.96 = $999.22

Page 19 — Centipede Subtraction (Subtracting whole numbers)

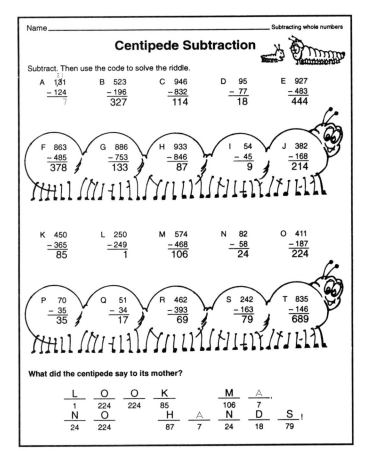

Subtract. Then use the code to solve the riddle.

A. 131 − 124 = 7
B. 523 − 196 = 327
C. 946 − 832 = 114
D. 95 − 77 = 18
E. 927 − 483 = 444
F. 863 − 485 = 378
G. 886 − 753 = 133
H. 933 − 846 = 87
I. 54 − 45 = 9
J. 382 − 168 = 214
K. 450 − 365 = 85
L. 250 − 249 = 1
M. 574 − 468 = 106
N. 82 − 58 = 24
O. 411 − 187 = 224
P. 70 − 35 = 35
Q. 51 − 34 = 17
R. 462 − 393 = 69
S. 242 − 163 = 79
T. 835 − 146 = 689

What did the centipede say to its mother?

L O O K   M A,
1 224 224 85   106 7

N O   H A N D S !
24 224   87 7 24 18 79

Page 20 — Missing in Action (Subtracting whole numbers)

Find the missing digits.

A. 5[2] − 39 = 13 ; 5 9 6 − 1 9 8 = 3 9 8 ; 6 2[1] − 1 1 [ ] = 6 1 0 ; 4 9 6 − 2 5[5] = 2 4 1

B. 8 3 − [4] 1 = 4 2 ; 4 5[0] − 2 6 5 = 1 8 5 ; 3 3 2 − [6] 7 = 2 6 5 ; 4 7 3 − [9] 2 = 3 8 1

C. 6 8 − 4[9] = 1 9 ; 9 1 − [6] 3 = 2 8 ; 7 7[4] − 1 7 = 7 5 7 ; 5[3] 2 − 1 6 6 = 3 6 6

D. 3 3[2] − 1 4 8 = 1 8 4 ; 3 5 3 − 1[7] 8 = 1 7 5 ; 8 9[5] − 1 6 7 = 7 2 8 ; 9 1 2 − 4 6[6] = 4 4 6

E. 2[5] 0 − 1 7 5 = 7 5 ; 9[0] − 3 6 = 5 4 ; 8 5 3 − [8] 4 = 7 6 9

F. 6[3] 3 − 3 7 = 6 2 6 ; 8 2 1 − 2[0] 6 = 6 1 5 ; 1 9[3] − 1 2 5 = 6 8

Answer Key

Page 21 — Leftovers
Subtracting whole numbers
Write the differences.

A. 5,237 − 1,856 = 3,381 | 6,895 − 2,167 = 4,728 | 1,596 − 898 = 698 | 6,735 − 1,948 = 4,787

B. 4,327 − 1,175 = 3,152 | 7,280 − 5,461 = 1,819 | 1,244 − 922 = 322 | 3,850 − 3,760 = 90

C. 5,556 − 2,658 = 2,898 | 4,392 − 1,774 = 2,618 | 6,432 − 265 = 6,167 | 5,324 − 1,427 = 3,897

D. 6,442 − 2,795 = 3,647 | 6,314 − 5,719 = 595 | 8,735 − 5,787 = 2,948 | 4,744 − 3,658 = 1,086

E. $42.38 − $9.39 = $32.99 | $17.71 − $2.88 = $14.83 | $92.63 − $65.94 = $26.69 | $15.35 − $13.69 = $1.66

F. $62.08 − $38.52 = $23.56 | $68.25 − $4.09 = $64.16 | $72.56 − $61.84 = $10.72 | $24.82 − $18.65 = $6.17

Page 22 — All That's Left
Subtracting whole numbers
Find the differences.

A. 4,529 − 1,635 = 2,894 | 6,218 − 3,862 = 2,356 | 9,126 − 7,241 = 1,885 | 7,843 − 3,589 = 4,254

B. 8,942 − 1,385 = 7,557 | 1,549 − 425 = 1,124 | 6,961 − 4,682 = 2,279 | 3,257 − 3,098 = 159

C. 8,573 − 2,791 = 5,782 | 9,836 − 1,465 = 8,371 | 3,816 − 942 = 2,874 | 8,414 − 3,916 = 4,498

D. 3,715 − 1,896 = 1,819 | 4,536 − 2,718 = 1,818 | 6,448 − 4,942 = 1,506 | 1,815 − 927 = 888

E. $15.95 − $9.98 = $5.97 | $53.10 − $49.95 = $3.15 | $87.50 − $49.25 = $38.25

F. $26.27 − $19.35 = $6.92 | $45.11 − $28.50 = $16.61 | $78.76 − $59.47 = $19.29

Page 23 — Across the River
Subtracting across zeros
Subtract.

A. 300 − 125 = 175 | 805 − 49 = 756 | 200 − 164 = 36

B. 502 − 159 = 343 | 4,307 − 439 = 3,868 | 5,900 − 1,375 = 4,525 | 5,302 − 1,953 = 3,349

C. 401 − 395 = 6 | 1,201 − 986 = 215 | 5,002 − 2,005 = 2,997 | 3,000 − 889 = 2,111

D. 6,004 − 4,832 = 1,172 | 5,089 − 498 = 4,591 | 5,024 − 1,954 = 3,070 | 4,702 − 1,794 = 2,908

E. $6.05 − $5.87 = $0.18 | $7.07 − $2.38 = $4.69 | $9.04 − $5.78 = $3.26 | $4.00 − $0.66 = $3.34

F. $58.05 − $19.98 = $38.07 | $20.05 − $9.87 = $10.18 | $31.06 − $13.67 = $17.39 | $80.05 − $76.19 = $3.86

Page 24 — Leapfrog
Subtracting across zeros
Subtract.

A. 300 − 174 = 126 | 100 − 53 = 47 | 402 − 394 = 8 | 700 − 492 = 208

B. 503 − 147 = 356 | 400 − 278 = 122 | 3,200 − 1,695 = 1,505 | 4,701 − 2,382 = 2,319

C. 1,702 − 375 = 1,327 | 2,903 − 1,594 = 1,309 | 6,001 − 5,873 = 128 | 4,067 − 1,970 = 2,097

D. 2,008 − 732 = 1,276 | 5,100 − 1,163 = 3,937 | 1,000 − 973 = 27 | 9,022 − 6,375 = 2,647

E. $8.01 − $2.32 = $5.69 | $3.01 − $0.67 = $2.34 | $80.40 − $23.63 = $56.77 | $14.06 − $5.98 = $8.08

F. $10.00 − $4.97 = $5.03 | $23.01 − $18.75 = $4.26

Answer Key

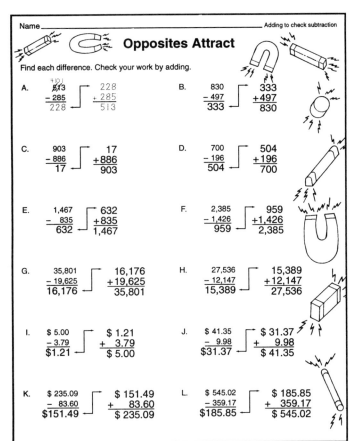

Page 25

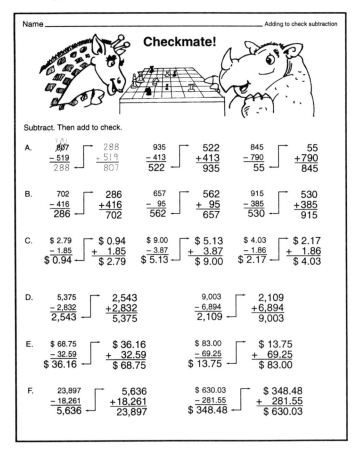

Page 26

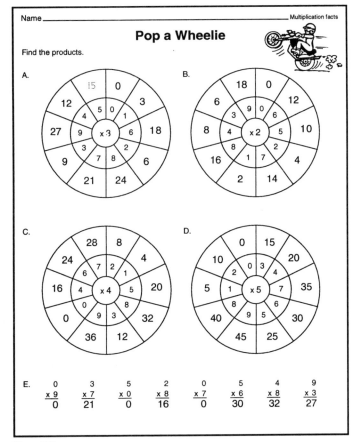

Page 27

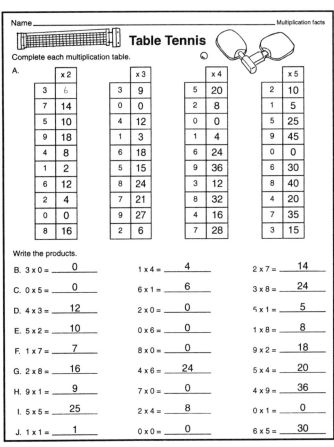

Page 28

111

Answer Key

Fast Facts — Page 29

Write the products.

A. 1 x 6 = 6
B. 4 x 7 = 28
C. 6 x 9 = 54
D. 2 x 7 = 14
E. 4 x 9 = 36
F. 8 x 8 = 64
G. 4 x 6 = 24
H. 7 x 4 = 28
I. 6 x 8 = 48
J. 1 x 8 = 8
K. 0 x 7 = 0
L. 6 x 6 = 36
M. 7 x 5 = 35
N. 7 x 7 = 49
O. 8 x 3 = 24
P. 9 x 9 = 81
Q. 9 x 6 = 54

5 x 8 = 40
2 x 6 = 12
7 x 8 = 56
3 x 6 = 18
4 x 8 = 32
0 x 9 = 0
2 x 9 = 18
9 x 4 = 36
3 x 8 = 24
0 x 6 = 0
7 x 3 = 21
8 x 4 = 32
7 x 6 = 42
8 x 9 = 72
6 x 7 = 42
8 x 6 = 48
9 x 5 = 45

3 x 9 = 27
5 x 7 = 35
5 x 9 = 45
1 x 9 = 9
7 x 9 = 63
9 x 8 = 72
9 x 3 = 27
9 x 7 = 63
0 x 8 = 0
5 x 6 = 30
8 x 5 = 40
3 x 7 = 21
2 x 8 = 16

Tune-up — Page 30

Write the products.

A. 5×6=30, 7×9=63, 8×8=64, 6×7=42
B. 1×7=7, 5×9=45, 6×8=48, 1×8=8, 4×7=28, 3×6=18
C. 2×9=18, 8×7=56, 4×6=24, 3×9=27, 8×6=48, 5×8=40
D. 3×6=18, 2×7=14, 3×8=24, 8×9=72, 7×6=42, 5×7=35
E. 7×7=49, 6×9=54, 7×8=56, 1×9=9, 9×7=63, 2×6=12
F. 3×7=21, 6×6=36, 4×8=32, 9×8=72, 9×6=54, 9×9=81

Fact Finder — Page 31

Circle groups of numbers that make multiplication facts. Look horizontally and vertically.

(grid of circled multiplication facts)

Product Power — Page 32

Complete the table.

X	1	7	5	9	8	2	6	0	3	4
6	6	42	30	54	48	12	36	0	18	24
8	8	56	40	72	64	16	48	0	24	32
0	0	0	0	0	0	0	0	0	0	0
3	3	21	15	27	24	6	18	0	9	12
1	1	7	5	9	8	2	6	0	3	4
7	7	49	35	63	56	14	42	0	21	28
9	9	63	45	81	72	18	54	0	27	36
5	5	35	25	45	40	10	30	0	15	20
4	4	28	20	36	32	8	24	0	12	16
2	2	14	10	18	16	4	12	0	6	8

© Frank Schaffer Publications, Inc. FS-32071 Fourth Grade Math Review

Answer Key

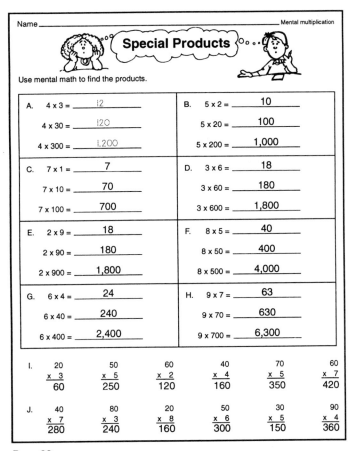

Page 33

Page 34

Page 35

Page 36

113

FS-32071 Fourth Grade Math Review

Answer Key

Page 37 — What's the Story?

Find the products. Regroup when you need to.

A. 45 × 4 = 180; 48 × 3 = 144; 13 × 9 = 117; 56 × 3 = 168; 42 × 2 = 84

B. 75 × 3 = 225; 58 × 5 = 290; 75 × 6 = 450; 14 × 8 = 112; 23 × 7 = 161

C. 29 × 4 = 116; 81 × 6 = 486; 36 × 4 = 144; 57 × 7 = 399; 15 × 7 = 105

D. $0.84 × 8 = $6.72; $0.63 × 5 = $3.15; $0.92 × 4 = $3.68; $0.38 × 6 = $2.28; $0.21 × 9 = $1.89

E. $0.47 × 4 = $1.88; $0.54 × 4 = $2.16; $0.92 × 6 = $5.52; $0.37 × 5 = $1.85; $0.32 × 6 = $1.92

F. Write a story problem for 3 × 75. Then find the answer. 225
Problems will vary.

Page 38 — Target Practice

Multiply. Regroup when you need to.
Circle the products that would round to 200 or $2.00.

A. 96 × 3 = 288; 23 × 3 = 69; 52 × 3 = (156); 43 × 6 = 258; 75 × 4 = 300

B. 36 × 4 = 144; 64 × 5 = 320; 25 × 7 = (175); 86 × 3 = 258; 18 × 6 = 108

C. 45 × 6 = 270; 58 × 4 = (232); 72 × 8 = 576; 94 × 7 = 658; 31 × 6 = (186)

D. 89 × 9 = 801; 68 × 2 = 136; 47 × 3 = 141; 29 × 4 = 116; 62 × 5 = 310

E. $0.78 × 5 = $3.90; $0.67 × 4 = $2.68; $0.28 × 7 = ($1.96); $0.43 × 9 = $3.87; $0.83 × 3 = ($2.49)

F. $0.93 × 3 = $2.79; $0.53 × 8 = $4.24; $0.84 × 6 = $5.04; $0.27 × 6 = ($1.62); $0.43 × 8 = $3.44

Page 39 — Join the Parade

Multiply.

A. 164 × 2 = 328; 412 × 3 = 1,236; 172 × 4 = 688; 198 × 2 = 396; 958 × 4 = 3,832

B. 426 × 5 = 2,130; 583 × 6 = 3,498; 395 × 3 = 1,185; 108 × 9 = 972; 282 × 3 = 846

C. 501 × 6 = 3,006; 362 × 2 = 724; 614 × 3 = 1,842; 492 × 2 = 984; 354 × 4 = 1,416

D. 347 × 5 = 1,735; 461 × 6 = 2,766; 927 × 3 = 2,781; 519 × 7 = 3,633; 863 × 8 = 6,904

E. $2.14 × 3 = $6.42; $5.07 × 3 = $15.21; $1.61 × 5 = $8.05; $2.83 × 8 = $22.64; $6.95 × 6 = $41.70

F. $1.26 × 9 = $11.34; $3.49 × 5 = $17.45; $6.12 × 4 = $24.48; $3.16 × 3 = $9.48; $2.08 × 3 = $6.24

Page 40 — It's What's Inside That Counts

Multiply. Then solve the riddle.

1. 315 × 4 = 1,260 (A); 206 × 9 = 1,854 (D); 344 × 7 = 2,408 (F); 746 × 4 = 2,984 (J); 376 × 2 = 752 (P)

2. 941 × 5 = 4,705 (G); 503 × 7 = 3,521 (M); 626 × 5 = 3,130 (W); 328 × 7 = 2,296 (O); 642 × 3 = 1,926 (H)

3. 708 × 5 = 3,540 (R); 569 × 3 = 1,707 (L); 121 × 8 = 968 (T); 936 × 7 = 6,552 (V); 861 × 4 = 3,444 (S)

4. 473 × 2 = 946 (B); 613 × 4 = 2,452 (Y); 748 × 9 = 6,732 (N); 712 × 6 = 4,272 (C); 271 × 7 = 1,897 (E)

What 8-letter word has only 1 letter in it?

A N   E N V E L O P E
1,260 6,732   1,897 6,732 6,552 1,897 1,707 2,296 752 1,897

Answer Key

Picture-Perfect Products
Find each product.

A. 1,413 × 4 = 5,652; 7,738 × 2 = 15,476; 6,184 × 5 = 30,920

B. 2,023 × 6 = 12,138; 5,670 × 3 = 17,010; 2,994 × 8 = 23,952; 6,092 × 3 = 18,276

C. 1,232 × 4 = 4,928; 5,033 × 9 = 45,297; 2,943 × 5 = 14,715; 3,438 × 3 = 10,314

D. 2,886 × 8 = 23,088; 4,433 × 6 = 26,598; 1,709 × 4 = 6,836; 6,330 × 2 = 12,660

E. $46.06 × 5 = $230.30; $21.01 × 4 = $84.04; $51.23 × 7 = $358.61; $68.93 × 3 = $206.79

F. $41.05 × 6 = $246.30; $73.14 × 8 = $585.12; $43.28 × 4 = $173.12; $68.43 × 7 = $479.01

Page 41

Production Line
Find each product.

A. 4,792 × 4 = 19,168; 3,412 × 2 = 6,824; 3,017 × 6 = 18,102; 2,999 × 4 = 11,996

B. 2,419 × 2 = 4,838; 1,372 × 4 = 5,488; 4,953 × 2 = 9,906; 7,621 × 8 = 60,968

C. 7,420 × 6 = 44,520; 7,231 × 6 = 43,386; 2,095 × 8 = 16,760; 2,645 × 3 = 7,935

D. 2,018 × 5 = 10,090; 1,343 × 2 = 2,686; 1,917 × 4 = 7,668; 5,473 × 5 = 27,365

E. $23.45 × 9 = $211.05; $64.92 × 7 = $454.44; $19.95 × 5 = $99.75; $23.06 × 6 = $138.36

F. $75.55 × 3 = $226.65; $19.99 × 6 = $119.94

Page 42

Quotient Quiz
Divide.

A. 18 ÷ 3 = 6; 10 ÷ 2 = 5; 21 ÷ 3 = 7
B. 15 ÷ 5 = 3; 12 ÷ 3 = 4; 8 ÷ 2 = 4
C. 4 ÷ 2 = 2; 7 ÷ 1 = 7; 9 ÷ 3 = 3
D. 10 ÷ 5 = 2; 6 ÷ 2 = 3; 15 ÷ 3 = 5
E. 24 ÷ 4 = 6; 18 ÷ 2 = 9; 30 ÷ 5 = 6
F. 4 ÷ 4 = 1; 16 ÷ 4 = 4; 8 ÷ 4 = 2
G. 45 ÷ 5 = 9; 12 ÷ 2 = 6; 3 ÷ 3 = 1
H. 5 ÷ 1 = 5; 16 ÷ 2 = 8; 36 ÷ 4 = 9

I. 1)6 = 6; 2)10 = 5; 3)6 = 2; 4)12 = 3; 5)5 = 1
J. 3)3 = 1; 5)40 = 8; 4)32 = 8; 2)18 = 9; 1)1 = 1
K. 4)28 = 7; 5)35 = 7; 3)18 = 6
L. 5)25 = 5; 4)20 = 5; 4)24 = 6

Page 43

Driving Division
Write the quotients.

A. 4)24 = 6; 2)6 = 3; 3)12 = 4; 4)4 = 1; 1)5 = 5
B. 3)18 = 6; 2)16 = 8; 3)3 = 1; 1)6 = 6; 5)10 = 2
C. 4)8 = 2; 3)6 = 2; 5)15 = 3; 3)24 = 8; 3)9 = 3
D. 2)4 = 2; 5)5 = 1; 2)8 = 4; 1)1 = 1; 4)12 = 3

E. 12 ÷ 3 = 4; 16 ÷ 4 = 4; 36 ÷ 4 = 9
F. 18 ÷ 2 = 9; 8 ÷ 1 = 8; 10 ÷ 2 = 5
G. 35 ÷ 5 = 7; 14 ÷ 2 = 7; 25 ÷ 5 = 5
H. 15 ÷ 3 = 5; 32 ÷ 4 = 8; 7 ÷ 7 = 1
I. 24 ÷ 4 = 6; 20 ÷ 5 = 4; 28 ÷ 4 = 7
J. 45 ÷ 5 = 9; 12 ÷ 2 = 6; 21 ÷ 3 = 7
K. 20 ÷ 4 = 5; 30 ÷ 5 = 6; 4 ÷ 1 = 4
L. 27 ÷ 3 = 9; 2 ÷ 2 = 1; 40 ÷ 5 = 8

Page 44

© Frank Schaffer Publications, Inc.   FS-32071 Fourth Grade Math Review

Answer Key

Page 45 — Pattern Paths

Write the quotients for each row. Look for a pattern. Then write a division fact that continues the pattern.

A. 6)12 = 2; 6)18 = 3; 6)24 = 4; 6)30 = 5; 6)36 = 6

B. 5)30 = 6; 6)36 = 6; 7)42 = 6; 8)48 = 6; 9)54 = 6

C. 7)63 = 9; 7)56 = 8; 7)49 = 7; 7)42 = 6; 7)35 = 5

D. 9)81 = 9; 8)72 = 9; 7)63 = 9; 6)54 = 9; 5)45 = 9

E. 3)24 = 8; 4)32 = 8; 5)40 = 8; 6)48 = 8; 7)56 = 8

F. 9)63 = 7; 8)56 = 7; 7)49 = 7; 6)42 = 7; 5)35 = 7

G. 8)72 = 9; 8)64 = 8; 8)56 = 7; 8)48 = 6; 8)40 = 5

H. 5)15 = 3; 6)18 = 3; 7)21 = 3; 8)24 = 3; 9)27 = 3

Page 46 — Marble Mania

A. 18 ÷ 6 = 3; 27 ÷ 9 = 3; 40 ÷ 8 = 5
B. 14 ÷ 7 = 2; 8 ÷ 8 = 1; 30 ÷ 6 = 5
C. 36 ÷ 6 = 6; 6 ÷ 6 = 1; 32 ÷ 8 = 4
D. 28 ÷ 7 = 4; 12 ÷ 6 = 2; 7 ÷ 7 = 1
E. 42 ÷ 7 = 6; 21 ÷ 7 = 3; 45 ÷ 9 = 5
F. 9 ÷ 9 = 1; 36 ÷ 9 = 4; 24 ÷ 8 = 3
G. 24 ÷ 6 = 4; 54 ÷ 9 = 6; 35 ÷ 7 = 5
H. 16 ÷ 8 = 2; 18 ÷ 9 = 2; 48 ÷ 8 = 6

I. 9)63 = 7; 6)48 = 8; 7)56 = 8; 9)81 = 9; 9)54 = 6
J. 8)64 = 8; 7)42 = 6; 6)42 = 7; 7)63 = 9; 8)72 = 9
K. 6)54 = 9; 7)49 = 7; 8)56 = 7; 9)72 = 8; 7)35 = 5

Page 47 — A Rule to Live By

#					
1.	T 9)9 = 1	A 6)30 = 5	T 2)2 = 1		
2.	F 9)81 = 9	R 8)56 = 7	E 9)18 = 2	H 7)42 = 6	
3.	A 7)35 = 5	F 3)27 = 9	R 9)63 = 7	C 9)36 = 4	
4.	E 8)16 = 2	K 6)48 = 8	E 7)14 = 2	O 8)24 = 3	
5.	H 6)36 = 6	A 9)45 = 5	K 5)40 = 8	O 3)9 = 3	
6.	F 6)54 = 9	R 7)49 = 7	A 8)40 = 5	H 8)48 = 6	

Write the letters on the lines to find a good rule to follow.

T A K E   C A R E   O F
1 5 8 2   4 5 7 2   3 9

T H E   E A R T H
1 6 2   2 5 7 1 6

Page 48 — Wishful Thinking

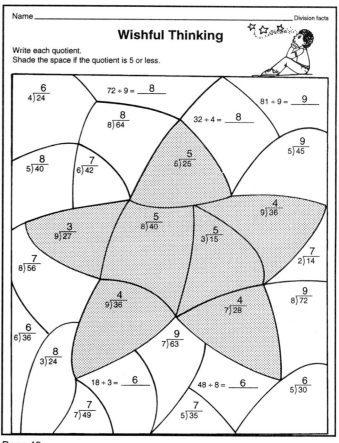

72 ÷ 9 = 8; 81 ÷ 9 = 9; 32 ÷ 4 = 8; 18 ÷ 3 = 6; 48 ÷ 8 = 6

4)24 = 6; 8)64 = 8; 5)40 = 8; 6)42 = 7; 5)25 = 5; 5)45 = 9; 9)36 = 4; 9)27 = 3; 8)40 = 5; 3)15 = 5; 2)14 = 7; 8)56 = 7; 9)36 = 4; 7)28 = 4; 9)72 = 8; 6)36 = 6; 3)24 = 8; 7)63 = 9; 5)30 = 6; 7)49 = 7; 5)35 = 7

Answer Key

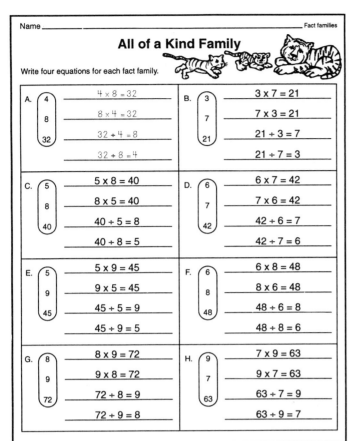

Page 49

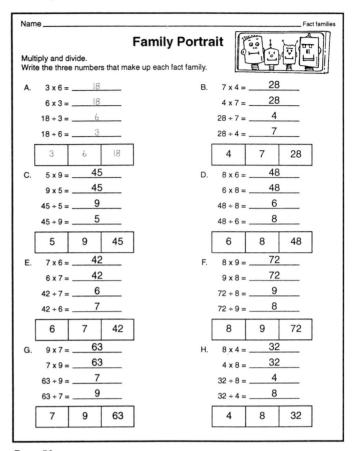

Page 50

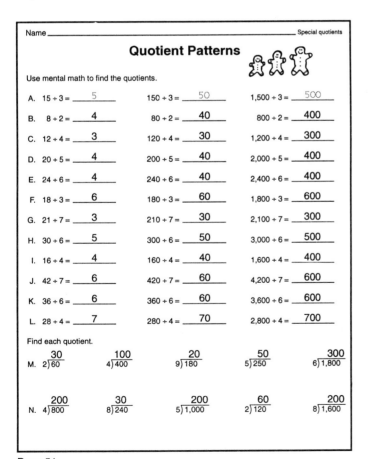

Page 51

Page 52

117 FS-32071 Fourth Grade Math Review

© Frank Schaffer Publications, Inc.

Answer Key

Friendly Numbers
estimating quotients

Use compatible numbers to estimate each quotient. Compatible numbers are combinations of numbers that are easy to divide mentally.

322 is close to 320. Since 32 ÷ 4 = 8, 320 ÷ 4 = 80

A. 322 ÷ 4 =
   320 ÷ 4 = 80

B. 123 ÷ 3 =   147 ÷ 5 =
   120 ÷ 3 = 40   150 ÷ 5 = 30

C. $2.41 ÷ 6 = $6.31 ÷ 7 =
   $2.40 ÷ 6 = $0.40 $6.30 ÷ 7 = $0.90

D. 117 ÷ 4 =   267 ÷ 9 =
   120 ÷ 4 = 30   270 ÷ 9 = 30

E. $3.63 ÷ 9 = $4.51 ÷ 5 =
   $3.60 ÷ 9 = $0.40 $4.50 ÷ 5 = $0.90

F. 103 ÷ 5 =   237 ÷ 8 =
   100 ÷ 5 = 20   240 ÷ 8 = 30

G. $4.19 ÷ 6 = $5.43 ÷ 6 =
   $4.20 ÷ 6 = $0.70 $5.40 ÷ 6 = $0.90

H. 477 ÷ 8 =   143 ÷ 2 =
   480 ÷ 8 = 60   140 ÷ 2 = 70

I. $8.07 ÷ 9 =   183 ÷ 3 =
   $8.10 ÷ 9 = $0.90   180 ÷ 3 = 60

Page 53

Over or Under
estimating quotients

Use compatible numbers to estimate each quotient. Compatible numbers are combinations of numbers that are easy to divide mentally.

628 is close to 630. Since 63 ÷ 9 = 7, 630 ÷ 9 = 70

A. 628 ÷ 9 =   83 ÷ 4 =
   630 ÷ 9 = 70   80 ÷ 4 = 20

B. 318 ÷ 8 =   477 ÷ 8 =
   320 ÷ 8 = 40   480 ÷ 8 = 60

C. 244 ÷ 3 =   209 ÷ 7 =
   240 ÷ 3 = 80   210 ÷ 7 = 30

D. $8.96 ÷ 3 = $4.19 ÷ 7 =
   $9.00 ÷ 3 = $3.00 $4.20 ÷ 7 = $0.60

E. 149 ÷ 5 =   723 ÷ 9 =
   150 ÷ 5 = 30   720 ÷ 9 = 80

F. $3.96 ÷ 8 = $8.11 ÷ 9 =
   $4.00 ÷ 8 = $0.50 $8.10 ÷ 9 = $0.90

G. 643 ÷ 8 =   492 ÷ 7 =
   640 ÷ 8 = 80   490 ÷ 7 = 70

H. 358 ÷ 9 =   324 ÷ 4 =
   360 ÷ 9 = 40   320 ÷ 4 = 80

I. $5.61 ÷ 7 = $5.39 ÷ 9 =
   $5.60 ÷ 7 = $0.80 $5.40 ÷ 9 = $0.60

Page 54

Snow Fun
Dividing 2 digits by 1 digit

Divide. Each quotient has a remainder.

A. 5 R 2 / 4)22 −20 / 2   6 R 2 / 4)26   8 R 2 / 6)50   6 R 1 / 2)13

B. 4 R 2 / 3)14   4 R 3 / 5)23   4 R 3 / 4)19   2 R 3 / 7)17

C. 7 R 3 / 4)31   3 R 2 / 7)23   6 R 1 / 3)19   8 R 3 / 6)51

D. 6 R 1 / 6)37   8 R 3 / 8)67   1 R 6 / 9)15   7 R 4 / 5)39

E. 9 R 1 / 7)64   6 R 5 / 9)59

Page 55

Picnic Planning
Dividing 2 digits by 1 digit

Divide. Each quotient has a remainder.

A. 4 R 2 / 5)22 −20 / 2   6 R 3 / 6)39   6 R 3 / 5)33   9 R 4 / 8)76

B. 8 R 1 / 2)17   4 R 1 / 3)13   6 R 1 / 4)25   3 R 3 / 9)30

C. 8 R 1 / 8)65   5 R 1 / 2)11   4 R 3 / 6)27   7 R 4 / 7)53

D. 5 R 4 / 6)34   1 R 5 / 7)12   6 R 4 / 8)52   7 R 5 / 9)68

E. 8 R 3 / 7)59   6 R 2 / 4)26   4 R 8 / 9)44   3 R 7 / 8)31

Page 56

© Frank Schaffer Publications, Inc.   118   FS-32071 Fourth Grade Math Review

Answer Key

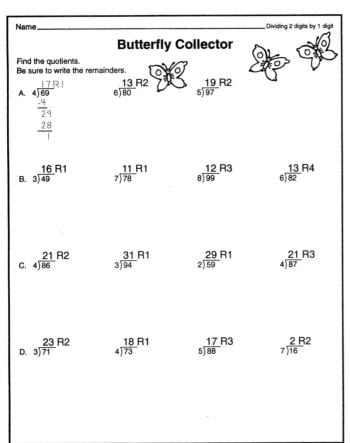

Page 57

Page 58

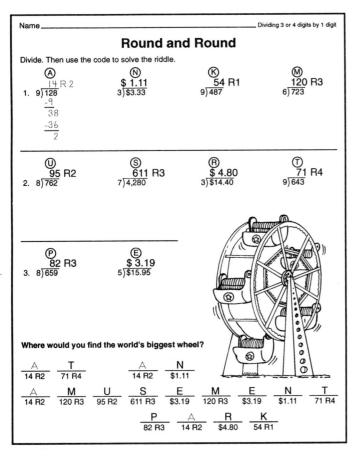

Page 59

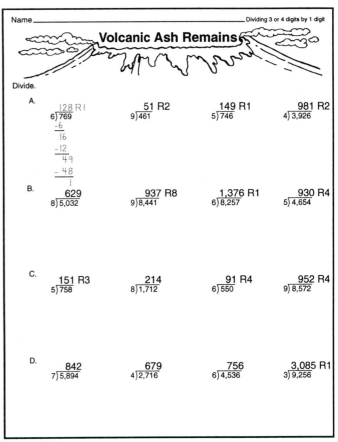

Page 60

119

Answer Key

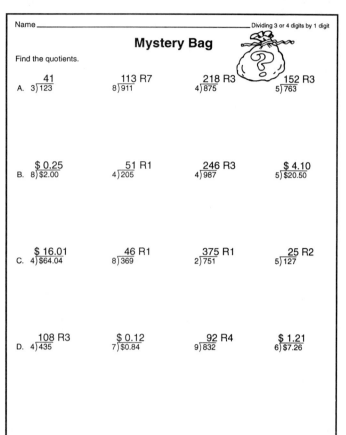

Page 61

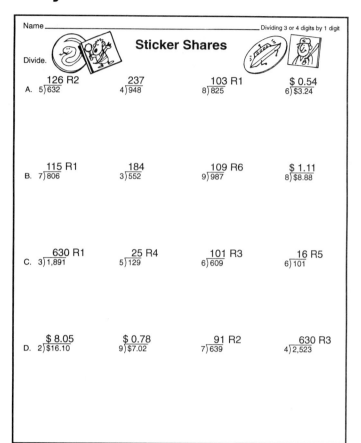

Page 62

Page 63

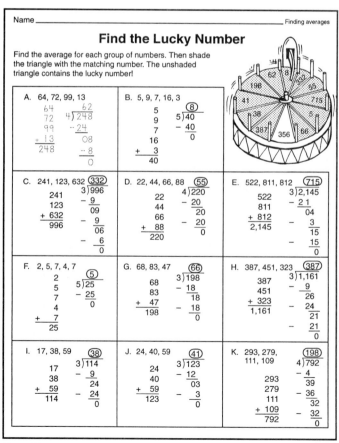

Page 64

120

Answer Key

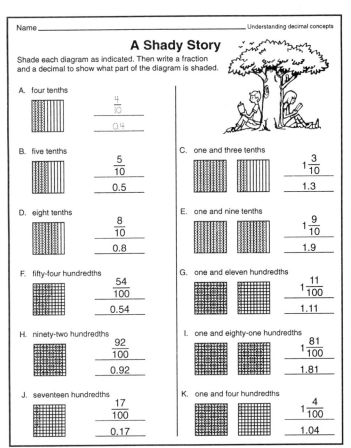

Page 65

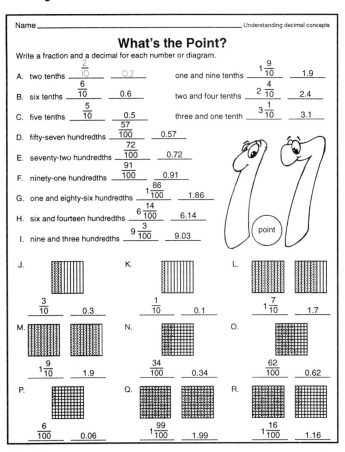

Page 66

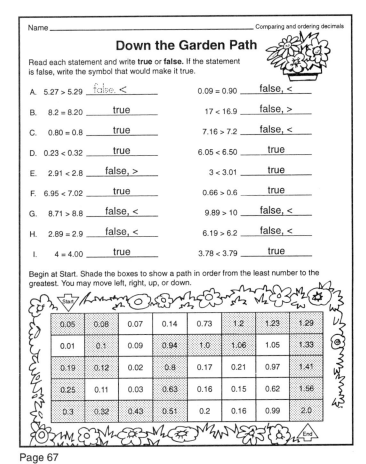

Page 67

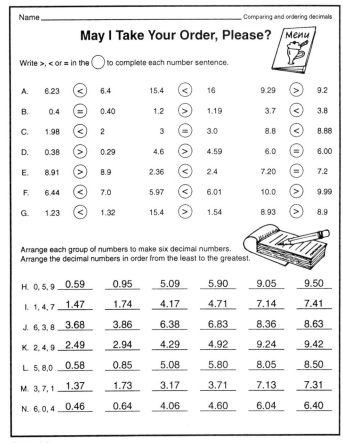

Page 68

Answer Key

Where's the Whole?
Round each decimal to the nearest whole number.

A. 3.4 __3__   4.8 __5__   8.2 __8__   3.7 __4__
B. 0.8 __1__   1.1 __1__   9.5 __10__   6.3 __6__
C. 16.2 __16__   25.3 __25__   84.9 __85__   93.6 __94__
D. 29.4 __29__   73.1 __73__   19.8 __20__   48.7 __49__
E. 3.72 __4__   4.88 __5__   9.14 __9__   5.63 __6__
F. 6.97 __7__   17.83 __18__   26.26 __26__   71.17 __71__
G. 49.65 __50__   99.51 __100__   2.71 __3__   16.61 __17__
H. 1.47 __1__   23.09 __23__   88.88 __89__   4.06 __4__
I. 25.11 __25__   2.55 __3__   46.36 __46__   39.10 __39__
J. 99.91 __100__   6.49 __6__   52.50 __53__   53.16 __53__

Look at the decimal numbers in the box below. Write each decimal number beside the whole number that it would round to.

54.8	53.75	55.25	52.95
53.26	54.5	54.05	53.19
55.43	53.49	53.99	53.87

K. 53 __53.26__ __53.49__ __52.95__ __53.19__
L. 54 __53.75__ __54.05__ __53.99__ __53.87__
M. 55 __54.8__ __55.43__ __54.5__ __55.25__

Page 69

The Whole Truth
Round each decimal number to the nearest whole number.

A. 2.9   3.2   0.6   6.7   8.1
   __3__ __3__ __1__ __7__ __8__

B. 71.3 16.4 94.5 29.6 31.3
   __71__ __16__ __95__ __30__ __31__

C. 81.6 43.5 58.2 46.7 96.8
   __82__ __44__ __58__ __47__ __97__

D. 2.73 3.64 4.13 8.88 4.95
   __3__ __4__ __4__ __9__ __5__

E. 16.82 27.28 99.51 19.62 36.49
   __17__ __27__ __100__ __20__ __36__

F. 12.65 63.14 52.83 76.09 15.34
   __13__ __63__ __53__ __76__ __15__

G. 80.92 27.77 55.68 29.83 10.44
   __81__ __28__ __56__ __30__ __10__

Write three numbers that round to each given number.

H. 10 __*Answers will vary.*__
I. 95 _____ _____ _____

Page 70

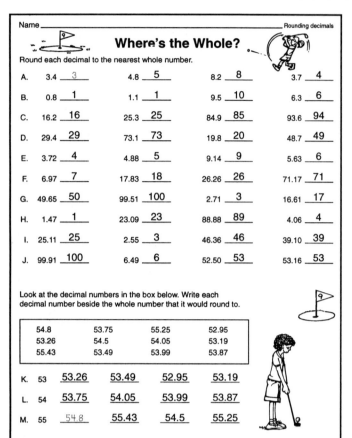
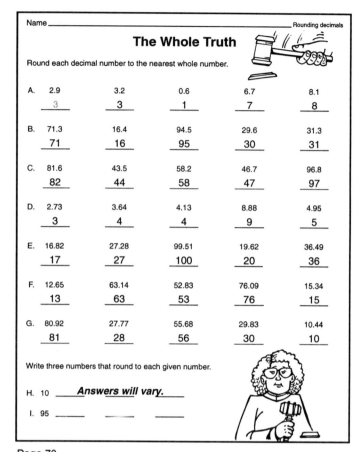

Ducky Decimals
Find the sums or differences. Watch the signs!

A.   3.4   45.3   8.4   40.8
   + 6.1   + 21.9   – 6.2   + 67.4
   ─────   ──────   ─────   ──────
    9.5   67.2   2.2   108.2

B.   7.7   83.3   55.6   71.6
   – 4.6   – 24.5   + 47.5   – 55.5
   ─────   ──────   ──────   ──────
    3.1   58.8   103.1   16.1

C. $ 53.70   0.88   43.75 $ 0.67
   + 85.68   – 0.48   + 82.19   + 0.81
   ───────   ──────   ───────   ──────
   $ 139.38   0.40   125.94 $ 1.48

D. $ 8.04   0.76 $ 1.79   62.83
   – 3.26   – 0.59   + 1.79   – 29.77
   ──────   ──────   ──────   ──────
   $ 4.78   0.17 $ 3.58   33.06

E. 51.0 – 6.4   82.61 – 14.48 $63.00 – 7.27
     51.0   82.61 $ 63.00
    – 6.4   – 14.48   –   7.27
    ─────   ──────   ────────
     44.6   68.13 $ 55.73

F. 48.06 + 3.41   4.09 – 0.79 $0.64 + $0.69
     48.06   4.09 $ 0.64
    + 3.41   – 0.79   + $ 0.69
    ─────   ─────   ────────
     51.47   3.30 $ 1.33

Page 71

Decimal Dots
Add or subtract. Watch the signs!

A.   6.3   7.8   15.45
   + 4.5   + 6.4   + 6.19
   ─────   ─────   ──────
    10.8   14.2   21.64

B. $ 7.02   64.2   17.6   0.72
   – 2.65   + 28.7   + 33.8   – 0.59
   ──────   ──────   ──────   ──────
   $ 4.37   92.9   51.4   0.13

C. $ 57.20   0.89   6.02   24.61
   – 4.84   + 0.53   – 0.69   + 16.18
   ──────   ──────   ──────   ──────
   $52.36   1.42   5.33   40.79

D. $ 35.09 $ 72.14 $ 36.05   60.47
   – 2.73   + 28.35   – 9.18   + 35.64
   ──────   ───────   ───────   ──────
   $ 32.36 $ 100.49 $ 26.87   96.11

E. 8.5 + 2.7   15.3 – 9.6 $5.31 – $0.82
      8.5   15.3 $ 5.31
    + 2.7   – 9.6   – $ 0.82
    ─────   ─────   ────────
     11.2   5.7 $ 4.49

F. 30.54 – 8.38   5.46 + 12.77 $35.19 + 23.81
      30.54   5.46 $ 35.19
    – 8.38   +12.77   + $ 23.81
    ──────   ─────   ────────
     22.16   18.23 $ 59.00

Page 72

© Frank Schaffer Publications, Inc.   122   FS-32071 Fourth Grade Math Review

Answer Key

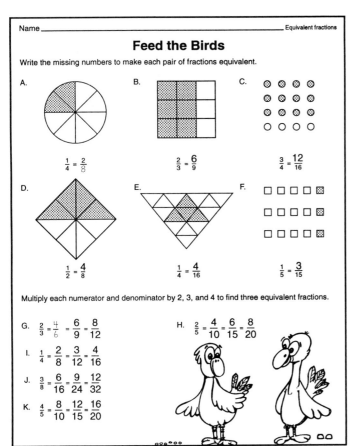

Page 73

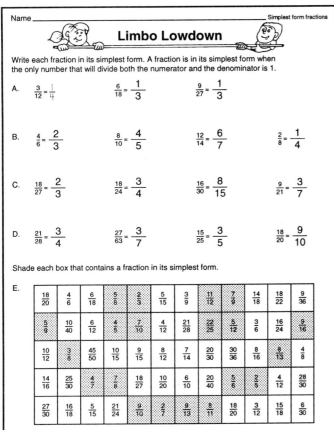

Page 75

Page 74

Page 76

123

Answer Key

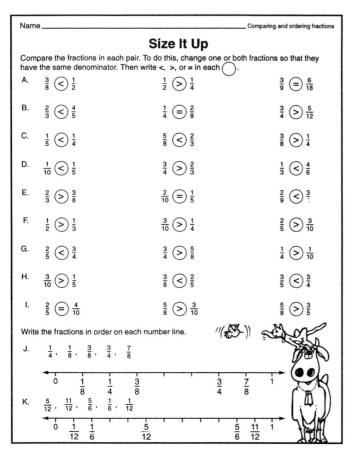

Page 77

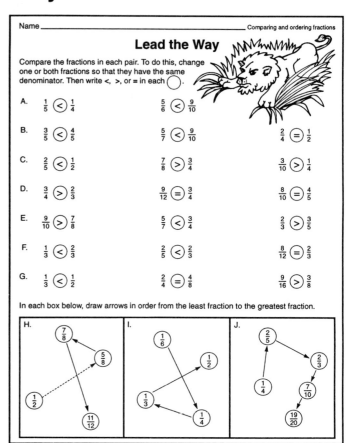

Page 78

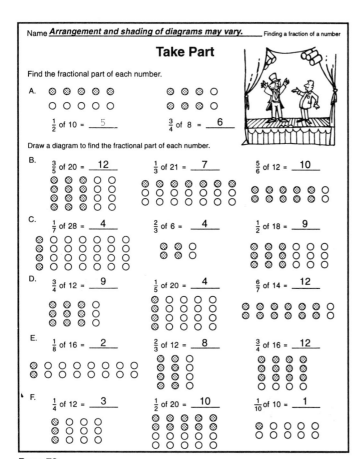

Page 79

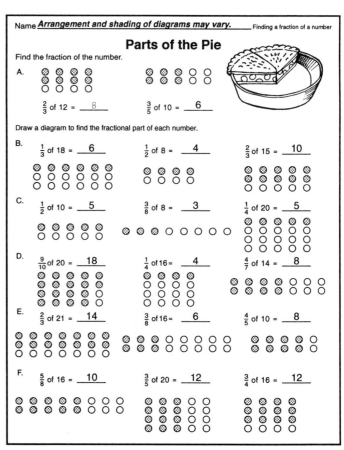

Page 80

Answer Key

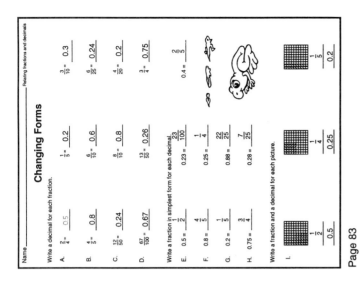

Page 83

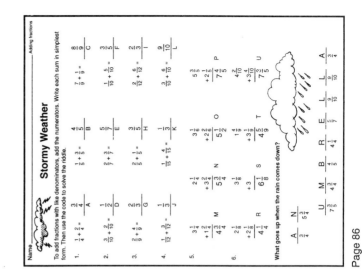

Page 86

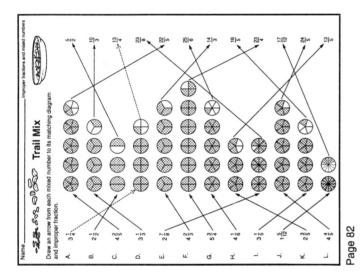

Page 82

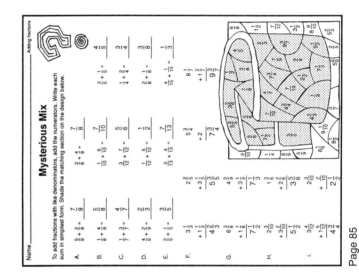

Page 85

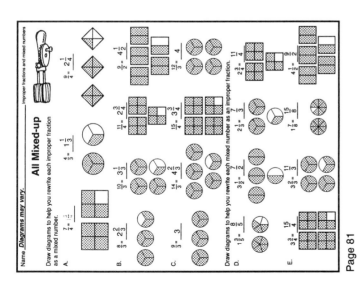

Page 81

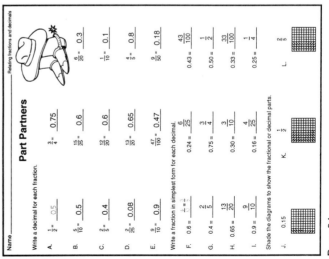

Page 84

Answer Key

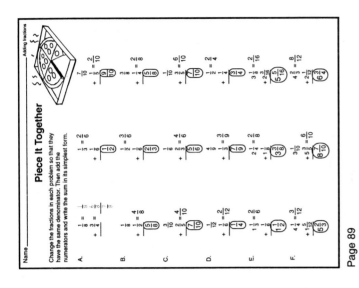

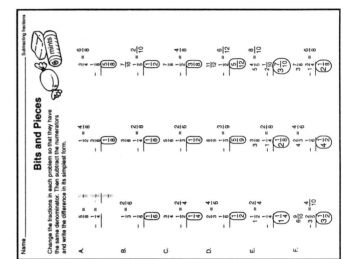

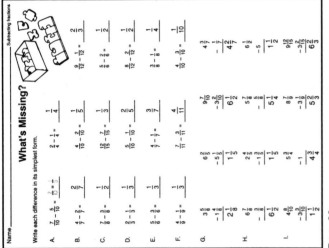

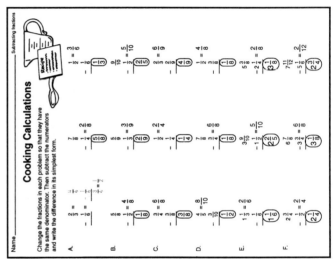

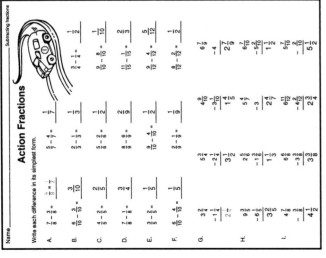

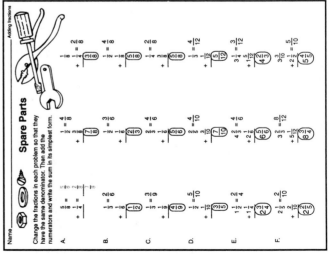

Answer Key

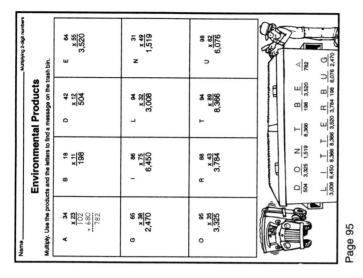

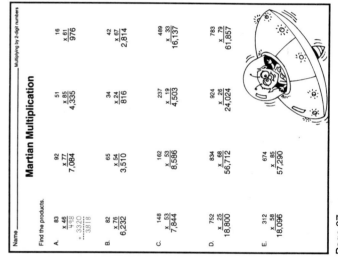

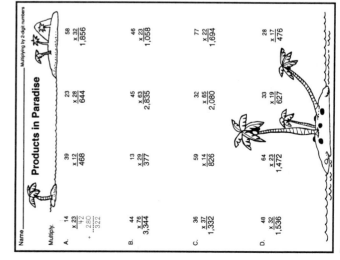

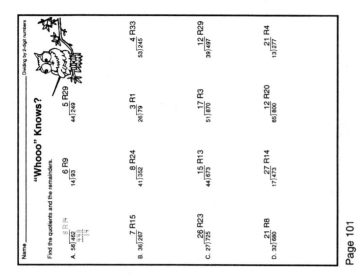